WILDFOWL

WILDFOWL

Photographs by Eric ^John^ Hosking

Text by Janet Kear

Foreword by Konrad Z Lorenz

Facts On File Publications
New York, New York • Bicester, England

First published in the United States by
Facts On File, Inc.
460 Park Avenue South
New York, New York 10016

Library of Congress Cataloging in Publication Data
Hosking, Eric John.
  Wildfowl.

  Includes index.
  1. Anatidae.   I. Kear, Janet.   II. Title.
QL696.A52H68 1985   598.4′1   84-21226
ISBN 0-8160-1152-4
Printed and bound in Singapore

**To Sir Peter Scott
the greatest Conservationist of our time**

Right. White-faced Whistling Ducks *Dendrocygna viduata.*

Jacket and title page illustration:
Grey Duck *Anas s. superciliosa* flying over New Zealand flax.

Contents

Foreword

by Konrad Z. Lorenz

Waterfowl are engaging creatures and that is probably the reason why more is known about their behaviour than about that of any comparable group of living animals. When the scientist uses the word 'amateur', or worse 'dilettante', he usually means it in a derogatory sense—and that is one of the most damaging fallacies. The error lies in the assumption that all science must be dry and boring, otherwise it lacks the dignity which is regarded as indispensable in making its statements respectable. This widespread misconception has been a curse of natural science for centuries. Even today there are people who despise descriptive science when it deals with objects of beauty. Workers so engaged are suspected of enjoying what they do. A brilliant physiologist of my acquaintance used to describe ornithologists as ornithomaniacs, albeit tolerantly and benevolently!

There is one science at least in which success is absolutely dependent on that non-rational joy which fixes the gaze of the investigator unflinchingly on the object of his research: that is ethology—the comparative study of behaviour. The existence of innate motor patterns and their similarities and dissimilarities, which enable us to construct the genealogy of living species, would never have been discovered if it had not been for the despised dilettantes and amateurs in general and ornithomaniacs in particular. The most important pioneers of comparative behaviour study, Charles Otis Whitman in America and Oskar Heinroth in Germany, were dedicated amateurs and bird-lovers—Whitman of pigeons, Heinroth of waterfowl. They discovered motor patterns which differed from species to species, from order to order and from family to family in exactly the same way and to the same degree as morphological characters do. In other words, they discovered the homology of motor patterns. This discovery is the archimedic point on which a comparative behavioural science could be constructed. The connotation of the term 'comparative' is exactly that of a science which from homologous characters in living organisms deducts the way in which they developed in phylogeny. What is generally described as comparative psychology is not comparative in the sense in which comparative anatomy is. Comparative ethology can only be successful in a taxonomic group containing many species, orders and families with a number of varied, graduated comparables, in other words, homologous characters.

I do not know of any group of animals which fulfils these conditions so completely as do waterfowl. Ernst Mayr and Jean Delacour have used ethological results for their taxonomy of Anatidae, Frank McKinney, Paul Johnsgard and other eminent scientists have enlarged our knowledge and made waterfowl a subject of evolutionary studies which become more important every year.

Motor patterns which are the subject of comparative ethology are extended in time and space. While the subject of comparative morphology is static and can be recorded in a still picture, motor patterns need the medium of cinematography. For representation in a book, photographs are a good substitute.

To give a satisfactory representation of a group like the Anatidae, verbal description and photography must go together. For the verbal description I could not imagine a better author than Janet Kear who has lived with waterfowl since early youth and has bred the taxonomically important forms such as Magpie Geese and screamers, thereby throwing some highly interesting sidelights on the evolution of the group. She has certainly made a very good job of rendering an overall impression in a compact and graphic manner. Her text and Eric Hosking's pictures fit together to form a satisfying whole. If any photographer is able, in a still picture, to convey a notion of what is happening, it is Eric Hosking.

The importance of the book is by no means confined to its ethological content; on the contrary, it conveys sound knowledge of the biology as well as the commercial importance of waterfowl and contains interesting historical facts concerning the domestication of ducks and geese. So the book will appeal to practical biologists as well.

But its main aim is indubitably the appeal to the sense of beauty which some people fortunately still retain. Man's interest in living creatures is always greater than that in a landscape or any inanimate object. The human interest aroused by higher animals and, especially, the emotional attraction of birds should make this book appeal to many people and should help to counterbalance the current estrangement from nature afflicting so many modern civilised people.

Acknowledgements

Overleaf. Martin Mere at sunset in November with roosting Pink-footed Geese *Anser brachyrhynchus*.

A party of American Wigeon *Anas americana*, Pintail *A. acuta* and Coot *Fulica americana* drinking in shallow water.

As with all of my previous books, I have been particularly fortunate in having an outstanding authority on each subject to write the text, and for this book there could be no one better than Dr Janet Kear to write about the world's wildfowl. She has lived and worked with them since her university days, is Sir Peter Scott's assistant as well as the Director of Martin Mere, the Lancashire wildfowl reserve. Furthermore, Professor Konrad Lorenz, Nobel Prize Winner and one of the greatest ethologists in the world has written the Foreword. I could not be in finer company. And who more appropriate to dedicate this book to than Peter Scott who has devoted so much of his life to the welfare and protection of the swans, geese and ducks of the world?

Once again I find myself greatly indebted to the people who have given me so much help in obtaining the photographs. We have travelled extensively to within 570 nautical miles of the North Pole and 800 miles of the South Pole to find them, and wherever we have travelled friends have done their utmost to make sure we achieved the best results. Dorothy, my wife, and David, our younger son, have been an enormous help and this book would not have come into being without their assistance. Dorothy now travels everywhere with me, carrying heavy photographic equipment and doing all she can to ensure I get the photographs. David has become an outstanding photographer in his own right and has supplied some of the pictures.

I cannot express enough thanks to Lars Eric Lindblad, Lars Wikander and their staff for inviting us aboard that unique ship *The Lindblad Explorer*. We were able to cruise to many parts of the world where no other passenger ship ventures, in order to take some of the photographs. More recently, Sven-Olof Lindblad of Special

Expeditions invited us to visit Alaska. While in New Zealand Ron Lockley and Geoff Moon took me to exactly the right locations to obtain wildfowl photographs, and in Kenya John Karmali, Peter Davey and John Williams were equally co-operative. In North America Ralph and Mildred Buchsbaum and Van and Peggy Van Sant were as keen as I was to help photograph new species. Back home in Britain Herbert Axell, the creator of the marvellous RSPB reserve at Minsmere knew the ideal places for me to get the best shots. Bill Makins, who owns a lovely part of Norfolk where he has created an ideal place for wildfowl, made us very welcome. Janet Kear not only wrote a splendid and informative text, but also invited Dorothy and me to stay so that I could take photographs of a few species in the Wildfowl Trust collection which I had been unable to obtain before.

As in my other books, I have included just a few black and white photographs taken by the late Niall Rankin while Dr D.P. Wilson willingly gave us permission to include his colour picture of the Ship's barnacles.

David would like to convey his thanks to Jane Dawson for giving his wife, Jean, and himself so much help while on Islay.

Photographs of the kind that appear in this book could never have been taken if we had not had the best possible photographic equipment and we are more than grateful to Barry Taylor and the Olympus Optical Company for supplying us with their cameras, lenses and other apparatus. Their new f/2.8—350mm lens is the finest we have ever used and there are several new items in production which will be of great benefit to the wildlife photographer.

We are delighted with the design of this book and thank Simon Blacker for all his work on it. We also owe much to our publisher, Croom Helm, and especially Christopher Helm and his Natural History Editor, Jo Hemmings, and their staff for all their expertise.

Finally we are thankful to the wildfowl themselves for the challenge they gave us in photographing them, which took us to some of the wildest and most beautiful parts of the world. If only man would protect their habitat and so help them to prosper, they could continue to give as much pleasure to others as they have given to us.

Introduction

I started to write this one evening in late autumn. Sitting at my desk, I look west through a picture window towards a rosy setting sun whose colours are reflected in 20 acres of shallow water. Outside are 100 Whooper Swans and 2,000 Pink-footed Geese from Iceland, plus 200 Bewick's Swans from Siberia and innumerable ducks. Three weeks ago Eric and Dorothy Hosking were sharing sight and sound with us; Eric was photographing and bubbling with enthusiasm and delight. What one could so easily forget is that the reserve, the lake and the house were all planned only a decade ago by Sir Peter Scott, to whom this book is dedicated. In 1972 there was little here but poor farmland, rough-grazed only in the summer, and a derelict farmhouse. With protection from disturbance, particularly from shooting, and the provision of permanent water and improved grassland, the wildfowl have come back to an area that they had left with the draining of the old Martin Mere nearly two centuries ago. But for Peter Scott's vision, I should not be here; without Eric Hosking's genius for catching the magic of ducks, geese and swans, the book would not have been possible, and without the early interest of Konrad Lorenz, who has written the Foreword, I might never have come to appreciate wildfowl anyway. For when I, as a student of animal behaviour at Cambridge, first met him, I was working on finches. The great man convinced me that, although all animals were worthy of study, he at least had been imprinted at an early age on ducks and geese and, for behavioural research, they were unsurpassed.

So this book brings together three of the men that I admire most, and a group of birds that I love and which, by the greatest of good fortune, provide me with a living. I have not written a complete thesis on the biology of wildfowl — the book is not long enough for that and there are many other publications that cover the subject thoroughly. It provides rather an introduction to the waterfowl group, a background for the lovely photographs and a selection of recent research that has interested me particularly. I have used in the main the classification of Paul Johnsgard, which was based on the masterly revision of the taxonomy of the family undertaken by Jean Delacour and Ernst Mayr in 1945. It is a pleasure to thank the following colleagues who corrected some of my errors: Andrew Dawnay, Denis Millington, Myrfyn Owen and Patrick Wisniewski. Lynda Seddon typed the manuscript (more than once!) and she and John Turner helped in many ways.

Janet Kear

CHAPTER 1

Evolution

Facing page. A pair of Magpie Geese *Anseranas semipalmata*. The gander, who is in front, has a rather higher crown to his head and is somewhat larger (he will weigh on average 2770 g. and his mate about 700 g. less) but otherwise male and female are indistinguishable.

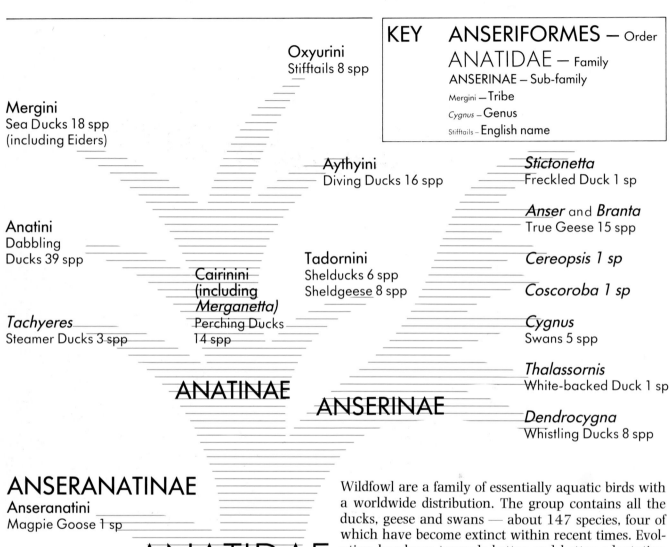

KEY   **ANSERIFORMES** — Order
ANATIDAE — Family
ANSERINAE — Sub-family
Mergini — Tribe
Cygnus – Genus
Stifftails – English name

Oxyurini
Stifftails 8 spp

Mergini
Sea Ducks 18 spp
(including Eiders)

Aythyini
Diving Ducks 16 spp

Stictonetta
Freckled Duck 1 sp

Anser and *Branta*
True Geese 15 spp

Anatini
Dabbling
Ducks 39 spp

Cereopsis 1 sp

Cairinini
(including
Merganetta)
Perching Ducks
14 spp

Tadornini
Shelducks 6 spp
Sheldgeese 8 spp

Coscoroba 1 sp

Tachyeres
Steamer Ducks 3 spp

Cygnus
Swans 5 spp

Thalassornis
White-backed Duck 1 sp

ANATINAE

ANSERINAE

Dendrocygna
Whistling Ducks 8 spp

ANSERANATINAE
Anseranatini
Magpie Goose 1 sp

ANATIDAE

ANHIMIDAE
(*Chauna*)
Screamers 3 spp

ANSERIFORMES

Wildfowl are a family of essentially aquatic birds with a worldwide distribution. The group contains all the ducks, geese and swans — about 147 species, four of which have become extinct within recent times. Evolution has been towards better and better adaptation to water-living, so that the birds at the 'bottom' of the family tree are essentially land birds that can swim, while those at the 'top' are water birds that only occasionally come ashore. It is assumed that the ancestors of the wildfowl group were truly land birds, as are the pheasants and game birds today. The screamers of South America are the closest to this ancestral type, and are the nearest relatives of the ducks, geese and swans, although quite different in appearance. Looking more like turkeys, they have longish legs with which they wade, huge, only slightly webbed feet, loose plumage and hooked hen-like bills. (More of the screamers later.)

If we compare the Magpie Goose, which has many features in common with the screamers, with the Musk

Duck, one of the stiff-tailed ducks at the 'top' of the wildfowl evolutionary tree, the consequences of the move from land to water will become more apparent. Both species are Australian, and both successful well-adapted marsh-living birds. To say that the Magpie Goose is primitive and the Musk Duck advanced must not be taken to mean that one is in some way 'better' than the other. In the struggle for existence, the fittest survive, but a later model may not necessarily fit an old habitat more adequately than the first.

The Magpie Goose *Anseranas semipalmata* is a rather untidy looking black and white bird, with a long neck and long legs. It can walk well on land; three of its toes (which point forward) are partially linked by webs, the other toe is long and points backwards at ground level.

(No bird has more than four toes — even *Archaeopteryx*, the first bird of which there is a fossil record, had only four — although the reptiles from which they are descended had five.) The plumage of the Magpie Goose is loose and the feathers are shed no more than once a year. The bird swims infrequently but can fly well and, when it moults its flight feathers, they are replaced in sequence like other land birds so that it is never flightless. Its bill is large, with a powerful 'nail', and it eats roots, herbage and the seeds of sedges and other marsh plants that it obtains by wading (it is almost entirely vegetarian). Goslings are hatched in a nest built for safety in the middle of the swamp, so that they need to swim more than their parents but still not a great deal, and they do not dive.

The Musk Duck *Biziura lobata*, by contrast, although also odd-looking to those whose typical wildfowl is a farmyard or park duck, is very well-designed for diving in water up to 6 m (20 feet) deep. It has smooth, tight-fitting plumage that is rather shiny like a grebe's, and it has great difficulty in walking on land because its legs are placed far back on the body where they are used as underwater paddles. The three forward toes are fully webbed and the hind toe or 'hallux' is tiny and lobed. Body feathers are moulted twice a year (they need to be thoroughly waterproof and presumably wear out more quickly than do those of land birds), and the wing feathers are moulted all at once, so that the bird is flightless for a while as the next set grows. The tail feathers are stiffened and used together as a rudder underwater; they may also be replaced more than once a year. The male has a large lobe or pouch that hangs below the bill like a wattle. The bill is stout, high and rather short, and the bird's diet consists at least partly of fish. The ducklings can dive when a day old and look like smaller versions of the adult.

As well as looking unalike, there also are great differences in the lifestyles of the Magpie Goose and the Musk Duck, and another evolutionary progression is obvious here. Some differences do not seem to be associated so much with the transition from land to water as with a common route taken by many bird groups: Magpie Geese have a stable family structure, often with more than one breeding female permanently mated to a single male, and where both sexes take an important part in nest building, incubation and the care of the young; the Musk Duck seems to have no pair bond, and the fertilised female is deserted soon after mating and left to bring up her offspring single-handed. This lack of paternal care in 'advanced' birds, with the females left holding the baby, is found in many other bird families as well as wildfowl.

As I have said, the Magpie Goose shows characters that are typical of evolutionary stages that came early in the development of the wildfowl and the Musk Duck has some that came late, but, although I may sometimes call these characters 'primitive' and 'advanced', this does not mean that either bird is any less or better adapted to an Australian marsh than the Common Shelduck *Tadorna tadorna* is to an Aberdeen estuary, or the Ruddy Duck *Oxyura jamaicensis* to Jamaica Bay. Both Magpie Goose and Musk Duck have fascinating 'peculiarities' that are not linked to any obvious evolutionary sequence, but are evidence of adaptation to a particular environment where these features enable the individual to compete successfully and to survive. Some of these 'peculiarities' will be mentioned later on.

As a group, wildfowl probably evolved in the swamp lands of the tropics and radiated from there. Their conspicuous success has been in colonising the north temperate and arctic regions. Northern areas have the largest number of wildfowl species and tropical regions the fewest. If we start at the bottom of the evolutionary tree on page 12, we come first to the screamers or Anhimidae.

Screamers will be considered at some length here, because of their special relationship to wildfowl, and then hardly referred to again. They bear little resemblance superficially to the ducks, geese and swans, but there are anatomical similarities that suggest that they form the link between the curassows, which are game birds, and that primitive-looking bird, the Magpie Goose. Like the Magpie Goose, their legs are long and their feet large with a strong hind toe and little webbing. On the other hand, the bill is rather game-bird-like, with a pronounced downward hook and none of the filtering fringes or 'lamellae' that are found in the bills of all ducks. Flight feathers are moulted gradually so that, like the Magpie Goose but unlike most other waterfowl, screamers do not pass through an annual flightless stage. A unique feature is that beneath the loose skin lies a network of small air-sacs, so that the bird crackles when it is handled.

Overleaf. A territorial encounter between two pairs of Shelduck *Tadorna tadorna* on the sands of the Ribble Estuary, Lancashire.

Facing page. Among the most aquatic of wildfowl, the male Musk Duck *Biziura lobata* weighs about 2400 g. and has a stiff tail for use as a rudder under water.

Below. A smaller stifftail, weighing only 510 g., the female Ruddy Duck *Oxyura jamaicensis* sits on her nest for 23 days.

There are three species, all native to northern South America. The Crested or Southern Screamer *Chauna cristata*, weighing up to 3–4 kg (8 lb), is the best known and is closely related to the slightly smaller Black-necked or Northern Screamer *C. torquata*, with which it has hybridised in captivity. Both species are grey, the Black-necked being darker with white cheeks and a longer crest. The tropical Horned Screamer *Anhima cornuta* is the largest, weighing 5 kg (11 lb), and is almost black above with white underparts. It has a remarkable long caruncle or horn (the function of which is unknown but probably connected with display) projecting from the forehead.

Screamers are mainly marshland birds, but are found also in open savannas and on the banks of ponds and slow-moving streams. Shallow water is used for roosting and they can swim, although they do so rather reluctantly. Chicks swim more frequently, especially while accompanying their wading parents. Once airborne, screamers fly strongly and can soar to considerable heights; they look even more like Magpie Geese when on the wing. The Crested Screamer grazes in open grassland along with farm stock, but it and the other species feed mainly on plants while wading through, or walking on, floating vegetation.

Male and female look alike and seem to pair for life. Courtship displays are inconspicuous, consisting mainly of duetting or antiphonal calling, rapid opening and closing of the bill, and mutual preening of the head and neck feathers. Pairs establish territories before breeding and flock together only in winter. Two sharp spurs protrude from the wrist of each wing and are present as tiny thorns at hatching; these are used to attack intruders, and broken-off spurs have been found buried in screamer breast muscle. All have loud, far-carrying voices with which they announce their possession of a territory and from which they derive their common name. The male's call is lower-pitched than the female's and is a double-noted *cha-ha*. They all also produce rumbling sounds by vibrating the air-sacs, which seem to function as a close-range threat display. A long breeding season is probably influenced by temperature and rainfall; nests are placed in shallow water within 80 m (86 yds) of the shore and built of sticks and vegetation. Material is not carried but passed back in the bill so that the nest is constructed of items within easy reach. Both sexes build, and both help incubate the three to seven large white eggs for 40–44 days. On hatching, the chicks are covered in dense yellow down and follow their parents from the nest. As in Magpie Geese, both parents may supplement feeding, and items are placed in the open gape of the chick; adults also pick up and drop food, apparently to bring it to their chicks' attention. Another unusual feature is that the young

are sometimes oiled from the parent's preen gland, which suggests that they are not especially well adapted to water-living.

After the screamers, the next species on the evolutionary tree is the first member proper of the wildfowl family, the Magpie Goose (which is so unusual that it has a separate subfamily to itself). Two other large

18

subfamilies make up the wildfowl, one of which (the Anatinae) can be called the duck subfamily and the other (Anserinae) the swan and goose group. English names are, however, sometimes confusing because 'duck' in common usage can mean any small waterfowl, while 'goose' is the title given to some members of all three subfamilies.

The Rushy Pen at Slimbridge in winter. The Bewick's Swans *Cygnus columbianus bewickii*, many ringed by Wildfowl Trust research workers, breed 2500 miles away in Siberia.

The Anserinae

Looking first at the Anserinae, the eight species of whistling duck *Dendrocygna* are probably the closest relatives of the Magpie Goose and, like that bird, are confined to the tropics and subtropics, although they have a much wider circumequatorial range. They are relatively long-legged and have an upright stance, moulting their body feathers only once a year. Male and female are alike and, as in the Magpie Goose, both sexes build the nest, incubate the eggs and care for the ducklings. Both sexes also utter a clear, often multi-syllabic whistle which is very distinctive and apparently important in keeping these somewhat gregarious birds together. Food tends to be plant material and a few species obtain this by diving in shallow water. Unlike swans and geese, the ducklings are distinctively patterned and, to the human eye, quite conspicuously striped — the black and yellow ducklings of the Red-

Above. The foot of a Mute Swan *C. olor* has three webbed toes plus a fourth smaller one pointing backwards. It supports a weight of about 10.5 kg.

Facing page. A 'southern' swan the Black-necked *Cygnus melanocoryphus* has short wings and often carries its cygnets on its back.

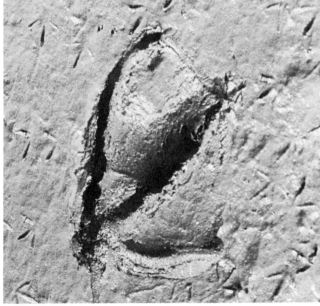

Left. The White-backed Duck *Thalassornis leuconotus* dives to feed on chironomid larvae.

Above. Footprint of a Mute Swan *C. olor* in the mud.

billed Whistling Duck *D. autumnalis* resembling downy wasps.

The diving White-backed Duck *Thalassornis leuconotus* of Africa and Madagascar is an oddity; at first glance, it is *very* well adapted to water, but nevertheless retains 'primitive' features, such as having males and females that look alike, male help in incubation and joint liability for the young. We do not know how often it moults its plumage; it does not have stiff tail feathers, and probably does not therefore use its tail very vigorously underwater. In many aspects of its display behaviour, and in its voice, it seems closest to the whistling ducks.

The swans, *Cygnus* and *Coscoroba* spp., are large —

the male Mute Swan *Cygnus olor* is among the heaviest flying birds. They have long necks and longish legs, and are good at walking but perhaps happiest floating on water. The 'southern' swans, from subtropical and temperate latitudes, comprise the Australian Black Swan *C. atratus*, its relative, the Mute, the Black-necked Swan *C. melanocoryphus* and the Coscoroba *Coscoroba coscoroba*, while the much more closely related group of four 'northern' swans consists of the large Whooper *Cygnus cygnus* and small Bewick's *C. columbianus bewickii* of Eurasia, and the large Trumpeter *C. buccinator* and smaller Whistler *C. c. columbianus* of North America. All male swans help to build the nest but only in the ap-

21

parently most primitive Black Swan (which is perhaps nearest to the ancestral swan) does the male also incubate regularly and for long periods. Males and females look alike and body feathers are moulted once a year. All swans have a vegetarian diet.

Cereopsis novaehollandiae, the Cape Barren Goose, is placed on different branches of the evolutionary tree by different authorities (as too is the White-backed Duck). It is a low temperate Australian species — so many Australian birds, like their mammal compatriots, are 'special' — the male and female are alike and the male helps build the nest (though he does not incubate). They are decidedly goose-like in appearance and feed by grazing but, as the small amount of webbing between their toes suggests, they seldom swim and change their body feathers only once a year. The youngsters are black and white and resemble young sheldgeese. Like young screamers, they are occasionally oiled by their parents, suggesting that they too are not all that well suited to the aquatic life.

The 15 species of 'true' grazing geese, ten grey geese *Anser* and five black geese *Branta*, occur only in the northern hemisphere. It has been suggested that they evolved from swans and that the fossil Maltese Swan *C. falconeri*, which had a swan-like body but short goose-like legs and feet, may have been an ancestor. As in swans, the male and female look alike but the male neither nest-builds nor incubates. Geese have one body moult a year, and walk well on land, but can also swim and regularly resort to lakes or estuaries to roost at night. Indeed many species go through the annual wing-moult on large stretches of water where they are safe from many predators.

Another Australian oddity, the Freckled Duck *Stictonetta naevosa*, is included with the swans and geese taxonomically although at first glance it looks much more like a dabbling duck because it has short legs. The scales on its legs are one of the clues that place it with the swans since they resemble those of 'primitive' wildfowl rather than those of the true ducks. It obviously has no very near relatives, and many aspects of its life history are still little known.

Pink-footed Geese *Anser brachyrhynchus* in Spitzbergen, during the annual wing-moult.

The Anatinae

Facing page. The Abyssinian Blue-winged Goose *Cyanochen cyanopterus* is a beautiful mountain sheldgoose, weighing about 1 800 g., still almost unknown in the wild.

We now come to the other great branch (or perhaps the main shoot) of wildfowl evolution, which has led to the ducks — the subfamily Anatinae. The sheldgeese and shelducks are at the lowest level on this branch and here it is often, though not always, possible to tell male and female apart by plumage, voice and behaviour. There are two moults of body feathers a year — at least in the shelducks, which seem to be more recently evolved than the sheldgeese and to be better adapted to the water. A double moult seems to be characteristic of groups that lead a more aquatic life. Paternal care is still assiduous when it concerns the ducklings but, as in the true geese and from here on up the evolutionary tree, the male has little or nothing to do with the nest and eggs. The young are generally black and white, although Kelp goslings *Chloëphaga hybrida* have pure white down and might therefore be said to look like young swans.

The sheldgeese are the grazing wildfowl of the southern hemisphere and are thought to belong to a more ancient group than the 'true' geese. Fossil sheldgeese occur in

Below. Flightless Steamer Ducks *Tachyeres brachypterus* in the sea off the Falkland Islands; the male on the left has the whiter head and is an aggressive bird. The short wings are not used for flight, but help in 'steaming' across the water surface.

both hemispheres, while bones of the *Anser* and *Branta* geese are found only north of the equator. This suggests that the sheldgeese may have become extinct in the northern regions in face of competition from the presumably better adapted 'true' grazing geese which eventually replaced them. The Blue-winged Goose of Abyssinia *Cyanochen cyanopterus* is the only grazing sheldgoose now occurring outside South America — indeed just into the northern hemisphere at 10° North. It is confined to the highlands above 2400 m (8000 feet) and can probably be regarded as remaining in remote mountain pastures where it is well adapted and has no competitors.

Steamer ducks *Tachyeres* of South America are related (but not very closely) to the shelducks and seem to be the southern hemisphere equivalents of the eiders *Somateria*, filling a similar coastal niche and feeding on molluscs and other marine invertebrates. They are good divers, and have two or even three plumage changes a year. Male and female are alike and both look after the young, although only the female incubates. There are four steamers, one very recently discovered, which look rather alike, and of which three are flightless. They are aggressive to their own kind, like the shelducks and sheldgeese, and highly territorial in defending a length of coastline.

The perching duck group contains a 'ragbag' of 14 species as unalike as the tiny pygmy geese *Nettapus* which eat lotus buds, the big aggressive Spurwinged Geese *Plectopterus gambensis* and the Torrent Ducks

25

Merganetta armata which are specialised for living in the cascading water of Andean streams. Primarily tropical and subtropical hole-nesters, in only a few species of this group is the male involved in parental care. This happens in the Brazilian Teal *Amazonetta brasiliensis*, Maned Goose *Chenonetta jubata*, Ringed Teal *Callonetta leucophrys*, Torrent Duck and in the Hartlaub's Duck *Pteronetta hartlaubi* but, interestingly, only in the last does the drake look like his wife. In contrast, the pair bond is short term in the Comb Duck *Sarkidiornis melanotos*, White-winged Wood Duck *Cairina scutulata*, Muscovy *C. moschata* and Spur-winged Goose, and seasonal in the North American Wood Duck or Carolina *Aix sponsa*, Mandarin *A. galericulata* and pygmy geese.

The Carolina and its relative the Asian Mandarin have very strikingly coloured drakes. Both sexes moult twice a year, so that the males can change into a drab,

Right. Despite its name the African Pygmy Goose *Nettapus auritus* is a small perching duck weighing only about 280 g.

Facing page. The rare White-winged Wood Duck *Cairina scutulata* from the rain forests of southeast Asia.

Right. Mother and duckling Carolina or North American Wood Duck *Aix sponsa* at the Laboratory of Ornithology, Cornell University, Ithaca.

Above. The largest perching 'duck', the male Spur-winged Goose *Plectropterus gambensis* weighs 6 kg.

female-like, well-camouflaged dress during the vulnerable period of the annual wing-moult and flightlessness. These two species are thought to be 'relics' of early radiations of perching ducks into the north temperate zones, a radiation that was replaced by the more recent evolution of the highly successful dabbling ducks.

The dabbling ducks are considered to be the most successful of wildfowl since, in numbers of species and in abundance, they outdo all the rest. 'Primitive' dabbling ducks tend to be tropical in distribution and to have dual parental care and sexes that look alike; however, such species are not numerous. More typical and widespread are the temperate forms that are adapted to marshy habitats and breed at latitudes north of those where they spend the winter. In these dabblers the male is brightly coloured during the period of courtship and dull brown during the late summer and early autumn which is the time that wing feathers are shed. Whenever the male has this 'eclipse' plumage (with one exception shown on page 29) he takes no part in family life. The drake Mallard *Anas platyrhynchos*, for example, is devoted to the same female for the season from October to April, but leaves her as soon as the clutch is laid, and may woo another wife in the following October.

In diving ducks also, the pair bond between male and female is seasonal, lasting only until she starts incubating her eggs. Again there is sexual dimorphism in plumage (male and female look different), so two moults of body feathers a year. However, male patterns tend to

27

be less complex than in the dabblers, and certainly there is no species with a particularly gorgeous drake. They are fresh water divers and mainly vegetarians, with legs placed well back on the body, large feet and a labed hallux. They have a worldwide distribution but have been particularly successful in the northern regions, where some migrate considerable distances within a continent between breeding and non-breeding locations.

The sea ducks (in which group I have included the eiders and the mergansers *Mergus*) are mostly salt water divers of which only one presently occurs in the southern hemisphere. They are sexually dimorphic in plumage, voice and display and they take (unlike most dabbling and diving ducks) two years to reach sexual maturity. The young are capable of diving at one day old and are well insulated with a layer of body fat beneath the skin, and a dense downy coat. The Long-tailed Duck *Clangula hyemalis* may have three moults of body feathers a year, since the drake has three plumage changes. Possibly some of the other marine divers replace their feathers fairly frequently. Male eiders are unusual in that, when they lose their beautiful breeding dress after their young

have hatched, they adopt a drab colouring for the wing-moult that is not brown and female-like. They tend to be blackish, and are still readily sexed even in eclipse.

The tropical and subtropical stifftails, which include the Musk Duck described in some detail at the beginning of the Chapter, are fresh water divers with legs placed so far back on the body that walking is difficult. The stiffened tail feathers are used to change direction while swimming beneath the surface. Presumably these tail feathers wear out quite rapidly and in most of the group wing and tail feathers are replaced more than once a year. The males frequently produce sounds not from the windpipe like other ducks but using inflatable air-sacs under the skin of the throat. They have elaborate displays, are usually unlike the females in appearance, and take no part in parental duties. Eggs are large, and from these hatch active, diving young which need relatively little parental care, even from the female. The only brood parasite among the ducks, the Black-headed Duck *Heteronetta atricapilla*, whose ducklings receive no attention from any adult, occurs in this group.

Above. A male Black-headed Duck *Heteronetta atricapilla* at Martin Mere. The female never makes a nest of her own but lays parasitically.

Facing page. A captive drake Long-tailed Duck or Oldsquaw *Clangula hyemalis*.

Right. The male New Zealand Brown Teal *Anas chlorotis* helps care for his young even when he is in eclipse plumage.

CHAPTER 2

Courtship

Courtship in birds, as in all animals, has evolved in order to procure a mate. The mate cannot be just anyone; he or she must be selected carefully so that age is correct, health and strength are good and, of course, the mate must be of the right species and the opposite sex. He or she must not be too closely related, but not too distantly either. With the best mate, an animal has an excellent chance of passing on its genes to the next generation.

Display is often the preliminary to choice of a mate, and visual and auditory displays may be highly conspicuous, involving the use of much colour and noise. In general in birds, it is the female that does the choosing and the males that provide the spectacle — in effect giving her plenty to select from. Display may also be involved in maintaining a pair bond that has been established already; if it is important for the survival of the offspring that the pair stay together, as in swans and geese for instance, then courtship may be seen throughout the lifetime of the partners, but it is often not very obvious and becomes less so as the birds grow old and, presumably, become more used to one another.

I have already mentioned in Chapter 1 some of the differences that exist in wildfowl 'marriage'. While Magpie Geese, whistling ducks and swans tend to pair for life with little divorce (especially in the migratory forms, such as the Bewick's Swan, where no case of divorce has been found during 20 years of study at Slimbridge), the more 'advanced' ducks, of which the north-temperate dabbling duck the Mallard is an example, tend to lose their mates at the end of every season and change their partners fairly regularly during their lives. These differences are associated with the amount of parental care that is given, and with the appearance of the two sexes.

Preceding page. Eiders *Somateria mollissima* in flight over Spitzbergen.

Above. Wild Mallard *Anas platyrhynchos* feeding on grain supplied in autumn at Martin Mere.

Facing page. A Whooper Swan pair *Cygnus cygnus* in Iceland. Apart from a slight difference in size — he weighs about 10.8 kg. and she 8.1 kg. — cob and pen are alike.

Dimorphism

Apart from a general dissimilarity in size, it is difficult to distinguish the sexes of any of the 'primitive' wildfowl on the right-hand branch of the evolutionary tree on page 12. It is hard to be sure whether a swan swimming alone is a male or female (sometimes called the 'cob' and 'pen'), and even an expert cannot tell whistling ducks apart simply by their appearance. The sheldgeese and shelducks are thought to be the modern representatives of an evolutionary stage that gave rise to the 'true' ducks, and they provide examples of both sexual dimorphism and plumage change during the year. If a bird has two annual plumages (a non-breeding and a breeding dress), then the obvious way to achieve the change is to moult body feathers twice. It was suggested in Chapter 1 that a double moult is one of the adaptations for water-living that appeared during the evolution of the progressively more aquatic ducks. (Auks and grebes are among the other water birds that also have two body moults and two annual changes in plumage pattern.) However, two species of sheldgoose are permanently dimorphic (the males being predominantly white and the females brown) and they may have only one change of feathers a year. So, if it is important for a sheldgoose that the sexes look different, the dissimilarity has to last for 12 months, not just for the breeding season. Their behaviour and voices are different also — the male is aggressive in display and in calls, and his main task is to defend the territory and to look conspicuous while doing so, so that rivals are in no doubt that the owner is in occupation. The female alone builds the nest and sits on the eggs and she is beautifully camouflaged for her task.

The shelducks, which probably evolved after the sheldgeese and appear to swim more, certainly have two moults a year, and the New Zealand Paradise Shelduck *Tadorna variegata* is unique in that it is the female that assumes a brightly coloured breeding dress — chestnut as opposed to the dark grey of the male — although she reverts to grey during the non-breeding season. She also has a permanently white head and there is a tendency in female shelducks of at least three other species to be somewhat more white-headed than the male. Why we do not know, but female shelducks appear to be more dominant than the male in other

A female Shelduck *Tadorna tadorna* and her half-grown young at Minsmere.

ways also — and they live to a greater age in captivity.

In most 'true' ducks, any difference between the sexes is the other way round — the male is the more conspicuous, and it is the drake that moults into a brightly-coloured breeding dress and a dull, 'eclipsed', female-like plumage in the non-breeding season. On average he lives to a greater age both in captivity and in the wild, and populations often contain a preponderance of males. An eye-catching drake is typical of the Carolina and Mandarin among the perching ducks, of most of the dabbling ducks, all the diving and sea ducks and most of the stifftails. And there is some tendency for the 'least tropical' to be the 'brightest'.

Voice and calls do not differ much between the sexes in the swans, geese and whistling duck group, while in shelducks and all the true ducks voices are very unalike. This results from great differences in the structure of the male's windpipe. Almost all female ducks just quack, and do so through a simple windpipe. 'Decrescendo calling', which has that characteristic falling cadence — QUACK, QUACK, quack, quack, quack — is an individual female's attempt to attract a mate and is a very common sound in any northern marsh in autumn and winter where the first decrescendo call is a sign that summer is passing. Each female seems to have her own way of calling and to call that way all her life, so that her mate may recognise and find her in the throng. The males' calls are much more variable from species to species. They utter in courtship a range of grunts, coos, whistles and honks that are often not very beautiful — although the cooing of the male eider is a most attractive noise. Some non-vocal sounds are also produced (quite commonly by the stifftails) in inflatable air-sacs near the throat, or by the whistle of the wings.

Facing page. The display of the male Torrent Duck *Merganetta armata*, on the right of the photograph, is spectacular. This is a captive pair and the male's tail feathers are very worn.

Right. King Eiders *Somateria spectabilis*. The males are in bright breeding plumage.

Below. This drake Carolina or North American Wood Duck *Aix sponsa* can be compared with the female's much less showy plumage on page 27.

Courtship

In species where the male and female look alike, display behaviour is often the only guide to an individual's sex, although courtship is not very different between male and female of the more 'primitive' wildfowl. In some, such as whistling ducks and the White-backed Duck, where calls are also similar, the only absolute behavioural difference is that males mount the females at copulation. In these birds, and in Magpie Geese, swans and true geese, pair-formation displays tend to be inconspicuous. The 'triumph ceremony' of swans and geese is an important element in maintaining the relationship of the pair. The male attacks or threatens other males or pairs (sometimes even an imaginary rival) and then returns to his mate displaying and calling, and she joins in with her head held low and close to his. In early courtship, the female may be reluctant to participate and pair-formation is not complete until she does so fully.

Among the sheldgeese, three species display with similar gestures from both sexes, but the rest and all the shelducks and 'true' ducks have displays that are

Left. During copulation, the male Egyptian Goose *Alopochen aegyptiacus* grasps feathers on the nape of the female's neck in order to maintain his balance.

Below. A displaying and noisy party of drake Wigeon *Anas penelope*. The females are out of the picture.

Right. A drake White-headed Duck *Oxyura leucocephala* in a 'head-high-tail-cock' posture displaying to a female.

Facing page. The 'head-throw' of a drake Common Goldeneye *Bucephala clangula.*

noticeably masculine. These male displays may be the result of females inciting their potential mates to attack other males, and selecting a final partner on the basis of his response.

While female displays tend to be rather similar from one duck to another, drakes may do many elaborately different things. In Mandarins and Carolinas the gaudy males display with movements that were originally used for preening and drinking, but over the course of evolution have been 'ritualised' to indicate something quite different to an observer and to show off large wing feathers or coloured plumes. Female dabbling, diving and sea ducks produce, by incitement, even more complex courtship postures from their males, and investigators have suggested that these include feeding, preening, bathing and aggressive components.

A group of courting Mallard drakes is a familiar sight in the temperate regions of the northern hemisphere during the autumn. They provide a showy water pageant, as their bright plumages are accentuated by displays and staccato calls. A single female showing an interest, by 'nod-swimming' towards the party, for instance, will produce a burst of activity. The three main postures are described by researchers in basic terms as 'grunt-whistle', 'down-up', and 'head-up-tail-up'. The female will select from the males one that will remain attentive to her all winter, especially while she feeds, will warn her of danger, and drive off drakes (who may attempt to rape her) and females from her nesting area.

The north European sea duck the Goldeneye *Bucephala clangula* also starts courting at low intensity in September although pair bonds are not usually formed until the end of the year. Sessions of display are seen in particular when a pair meets a lone male. There are a bewildering number of drake Goldeneye displays. One of the most common is called 'bowsprit-pumping': the neck is extended repeatedly along the water surface, and then withdrawn. The 'head-throw' is among the most conspicuous and is a rapid performance in which the head is tossed back to the rump and held there for a while as a rattling call is given.

The Mediterranean White-headed Duck *Oxyura leucocephala* is a stifftail in which the male differs from the brown female in having, in breeding condition, a bright blue bill much swollen at the base, a white head, chestnut breast, and grey and red-brown upper parts. The pair bond is brief, and the male displays to any female throughout the breeding season. She alone shows territorial aggression around her nest-site, and around her ducklings until they are about two weeks old.

Courtship

Sexual Maturity

The age of maturity — that age at which a duck, goose or swan will use courtship display to win a mate — is not the same in all wildfowl. In general (as in many animals) the larger the adult, the longer it takes to reach maturity. Swans may not breed until they are three or four years old; geese, the larger perching ducks and sea ducks not until they are two; while most of the smaller ducks can reproduce during the year following hatching.

It is typical in birds where there is an adolescent stage that a distinctive juvenile plumage or some other indication of immaturity is found. For instance, the juvenile swan has grey-brown feathers instead of white ones (even the young Black Swan is brown rather than black). The colour of the legs, eyes and bill in many ducks may intensify as adulthood is reached, or swellings near the head become enlarged and pigmented. These 'indicators', which are signals to other birds interested

Facing page. The incubating female Mute Swan *Cygnus olor* will sit for 36 days with breaks once or twice a day for feeding. During resting, the bill is often tucked into the feathers of the back so that warm air is drawn into the nostrils.

Below. A male Surf Scoter *Melanitta perspicillata*.

in acquiring a mate or in protecting an existing partner from unwelcome attention, are often more prominent in the male — and typically occur in the region of the head. A brown juvenile plumage is normal even in ducks that mature in a year. Male youngsters will moult into their first bright feathers only when they are old enough to court a female. In the 'true' ducks, however, most females continue to wear a juvenile dress all their lives. This phenomenon is called 'neotony' —reproduction occurs while the animal is still at what appears to be an adolescent stage. Hormones are probably responsible for maintaining the female's juvenile plumage, since, if her ovary is removed, an adult female will take on the bright drake pattern at the next moult. Sometimes disease may inhibit secretion from the ovary, and a female Mallard or Mandarin can be seen gradually to assume the male's coloration.

The importance of juvenile plumage in protecting the immature bird from hostility, since with this plumage they visibly present no competition for mates or territory, is well illustrated by the 'Polish' Mute Swan. Although Mute Swan cygnets are normally grey, individuals with white down occur from time to time. These birds pass straight into a white plumage, similar

to that of the adults, instead of into the usual grey-brown feathering. They also have pinkish legs and feet, instead of black ones, and retain these features into adult life. The pale-footed birds became known as Polish Swans among London poulterers, who used to import them from the Baltic. Indeed, about 20 per cent of Mutes in eastern Europe are 'Polish', while in western countries only 1 per cent are, although this figure seems to be increasing. The sex ratio of the white juveniles has been shown to be uneven — about three-quarters of them are female. Cygnets in white plumage lack the safeguard of looking adolescent and therefore harmless, and there are several records of birds being

Above. A flock of Mute Swans *Cygnus olor* in late summer at a good food source. Some of the birds are still moulting their wing and tail feathers.

Right. The introduced Canada Goose *Branta canadensis* in Britain is highly successful and increasing at about 8 per cent a year.

attacked by their parents and even killed. On one occasion, a female was seen trying to drown her white offspring; they reached the shore with difficulty where they were attacked by a crow. The male then rushed to their defence.

It is obviously surprising that the white phase occurs at all, except very rarely, since normal juvenile plumage has the advantage of preventing parental aggression, as well as attacks from other adults in the winter flocks. Can there be some compensating gain, especially to females in whom the sex-linked feature is so relatively common? It has been suggested that there is, and that this advantage results from the fact that 'Polish' females appear older than they really are. They can thus attract a male earlier than usual and so extend their breeding life span by a year. Wild male Mutes never pair in their second year whereas females sometimes do. Males seldom even breed in their third year, while females do so more often and more successfully. By appearing older than they are, female 'Polish' birds might obtain a mate in their first winter, spend their second (infertile) spring gaining experience of a partner and of a territory, and always breed successfully during the following year. Thus the 'Polish' sex-linked gene might confer a net profit despite parental aggression, at least in expanding populations breeding at low densities.

Finding a Partner

How does a bird know whom to court? How does a female duck, reared only by her mother, know that the very different-looking adult drake is the same species as she, and a potential father of her children? Male ducks seem to learn the plumage patterns of the opposite sex from the ducks with which they are reared. Females, on the other hand, seem able to ignore such early experience and to accept (e.g. by 'nod-swimming') the courtship of an appropriate male when they are mature. It is suggested that the drakes' plumage patterns are sufficiently bold and simple to be coded into the instinctive knowledge that the female inherits. We find that a single duckling of a dabbling duck reared by humans to the fledging stage may behave in two ways towards its human companions when it is mature. If it is a male, it will often direct its courtship display towards humans; a female will not. The reverse appears to be true of the Cape Barren Goose; females hand-reared alone continue to court humans, while males reared in a similar fashion

later mate normally with their own kind. Only in the dabbling ducks and *Cereopsis* do we know for certain that there is a clear distinction between the behaviour of the sexes; other wildfowl have not been studied to the same extent.

We know that many wildfowl 'imprint' on the first moving object that they see at hatching. At one time, it was thought that this process of very rapid learning later influenced the choice of a sexual partner. We now know that the situation is more complicated than that. Imprinting certainly ensures that individuals will recognise the species to which they belong (siblings are important here as well as the parent) and one of its many advantages may be that it ensures that a bird selects as a mate an animal sufficiently like its siblings to be of the right kind, but just sufficiently unlike its brothers and sisters to prevent inbreeding.

Below. Mallard ducklings *Anas platyrhynchos* will imprint on their mother and on one another as they hatch.

Facing page. A captive female Cape Barren Goose *Cereopsis novae-hollandiae*.

Pair Bonding

Courtship leads to a pair bond forming — and almost all wildfowl except stifftails and perhaps the Muscovy Duck are monogamous; although most of the 'advanced' ducks may change partners every season, males have only one wife at a time. An exception is again the Magpie Goose, where one male frequently establishes a stable relationship with more than one female, often apparently with a couple of sisters or the original mate plus a daughter. The sex ratio in Magpie Geese seems to favour the females and, in captivity at least, males die at a younger age, as is the case with shelducks.

An engagement period is common in ducks of the northern regions where the pair bond forms several months before breeding starts; in swans and geese there may be a year's engagement. Copulation itself may be an important part of maintaining the partnership between the sexes and be functional (in the sense that sperm is present to be passed) only just before the clutch is to be laid. In Magpie Geese, copulation occurs

at the nest site, as it does in many land birds; in some other relatively 'primitive' wildfowl such as the whistling ducks it happens while the birds are standing in shallow water; in the majority of species the male and female mate while swimming. Even the Hawaiian Goose *Branta sandvicensis*, whose habitat contains no permanent pools, goes through a display that consists of head-dipping and bathing movements before mating. The habit of mating on land is an example of the Hawaiian Goose's adaptation to an environment where water is scarce. The bathing movements, however,

Preceding pages. White-faced Whistling Duck *Dendrocygna viduata* at Lake Jipe in Africa.

Facing page. A stifftail from east and south Africa, a female Maccoa *Oxyura maccoa*.

Below. Compare the head of this Barnacle Goose *Branta leucopsis* with the Ship's Barnacles on page 144.

show that it must have evolved from an ancestor that in the distant past copulated while swimming. Such is not the case with the Cape Barren Goose: this bird also mates on land, but its display has no elements that suggest an ancestor more evolved for water-living than is the present-day species.

How long does sperm remain viable in the female's reproductive system after copulation? Can a single mating ensure that all the eggs of the clutch are fertilised? Sperm can be stored in the lining of the oviduct and, although we do not know the situation in many species, observations made after artificial insemination of domestic ducks suggest that one mating is sufficient. In geese that migrate, copulation may occur on the journey. Barnacle Geese *Branta leucopsis* mate in Iceland during their spring migration to Greenland, and start to lay almost as soon as they arrive at their final destination. There they may copulate throughout the laying period.

The lack of courtship feeding in the wildfowl group is notable. In only one duck, the Red-crested Pochard *Netta rufina*, does the male feed the female at all frequently, although the behaviour has also been reported in the Carolina. In the breeding season the drake Red-crested Pochard dives repeatedly for food and on surfacing shares a bill-full of weed with his mate who swims to him. In Chapter 4, I will consider both parental and courtship feeding, and the reasons why they may be rare in waterfowl.

Mutual preening is a little commoner than courtship feeding, especially in tropical species which have long pair bonds between sexes that are rather alike and have similar voices. It probably helps familiarise the pair with characteristics of one another that are important in recognition. Andean Geese *Chloëphaga melanoptera*, the sheldgoose where both sexes are white, preen one another in the region of the head, throat and face. So do most of the whistling ducks. Mutual preening is particularly common in the White-faced Whistling Duck *Dendrocygna viduata*, and it is to the patches of white on the face and neck that the nibbling action tends to be restricted. As well as having great social significance, the preening also helps to remove external parasites from regions that the bird cannot easily reach itself, and whistling ducks as a group actually lack a head louse that occurs on all other wildfowl.

A pair of White-faced Whistling Duck *Dendrocygna viduata*. The duck on the right is preening the head feathers of the left-hand bird whose eyes are half-shut.

CHAPTER 3

Nests and Eggs

What determines the timing of nesting and egg-laying in wildfowl? In the northern temperate regions, we are used to ducklings appearing on the ponds with their mother in the late spring and not at other times of the year. Domestic genes in many of our park Mallard do confuse the situation somewhat: domestic ducks have been bred selectively to lay over a very long season and occasionally ducklings hatch at entirely the wrong time — and usually perish unless rescued by Man. In the tropics the situation is different in that breeding takes place successfully over many months of the year.

Breeding seasons, like cycles of moult and fat deposition before migration, are physiologically controlled by the secretion of hormones. These secretions are regulated by seasonal changes among which are variations in length of day and night. The simplest egg-laying patterns exist among the Magpie Goose, whistling ducks and some 'primitive' perching and dabbling ducks of the tropics. Breeding begins when daylight is of approximately 12 hours' duration and continues until it falls below this level in the autumn; in consequence, the breeding season extends more or less symmetrically on either side of the summer solstice. In the wild, if the bird lives close to the equator, any relationship between light and laying is not obvious: the daylength never becomes much shorter or longer than 12 hours, so the nesting season is geared to other natural phenomena (such as rainfall) which influence the availability of food and nesting cover. But if these birds are moved out of the tropics, to the Wildfowl Trust at Slimbridge, for instance, then the effect of light becomes clearer. Captive Magpie Geese in England may lay at any time between April and September. The north temperate Pink-footed Goose *Anser brachyrhynchos* also starts breeding in April, but always stops doing so by midsummer. This is the typical pattern for the true geese and swans and, indeed, for all the north temperate ducks; the long days of high summer inhibit breeding and cause the birds to moult their wing feathers.

Above. Feeding the ducks at Waterlow Park, London N19.

Facing page. Plumed Whistling Ducks *Dendrocygna eytoni* and Pied Stilts *Himantopus leucocephalus* at an Australian marsh.

Some other Slimbridge nesting records are interesting. The temperate Cape Barren Goose from Australia lays between November and the end of March, right through and including the winter solstice. This situation is the opposite of that found in the more tropical Magpie Goose which also comes from Australia: short days, even as short as 7.7 hours of light, do not stop the Cape

Above. The incubating Long-tailed Duck *Clangula hyemalis* takes 26 days to hatch her eggs. Compare her drab appearance with that of the male on page 28.

Right. The female of the Hooded Merganser *Mergus cucullatus* nests in tree-holes in Canada and the United States. Here are two females and a male.

Below. The eggs of the Ruddy Duck *Oxyura jamaicensis* are large so that the duckling is plump and well-insulated at hatching.

Barren Goose from breeding, while long days do. In the wild, it comes into breeding condition in the winter when pastures will be lush enough to provide adequate rearing food for its goslings. High summer in Australia would be a very inauspicious time for a grazing bird to hatch.

It is found that most wildfowl require a specific day-length before they can lay. Actually, it is likely that it is the males whose gonads are more stimulated by the appropriate photoperiod, and that they stimulate the females by displaying. Ultimately, however, it is the food supply of the female and of the young that provides the important evolutionary pressure behind any differences in the time of wildfowl breeding.

The selection of a safe nesting site is of critical importance not only to the eggs but to the bird that must warm those eggs for a matter of three or four weeks. Nests built over water among marsh plants are fairly typical of 'primitive' wildfowl such as the Magpie Goose, and the safety from predators thus obtained may have been the evolutionary force that led the rest of the group towards a more aquatic existence. Many wildfowl nests, however, are built on dry land concealed by overhanging vegetation, and this is typical of dabbling and diving ducks. Some species, such as shelducks, use burrows or cavities under boulders, and others, like the perching ducks and most mergansers, use holes in trees. These burrows and tree holes are not constructed by the ducks; the shelduck certainly cannot dig a burrow nor a Goosander *Mergus merganser* fashion a hole even in a rotting tree trunk where the wood is soft. They take over what is available in such places as Rabbit (*Oryctol-*

agus cuniculus) warrens and trees inhabited by wood-peckers. Nevertheless, there are some structural adaptations needed for nesting in holes. Look at the claws of the Goosander female clinging to the bark of a tree in the photograph on this page. They are prominent and sharp; so too will be the claws of her ducklings, since they will need to climb out of the nest and join her on the water below. It is interesting to speculate on where British Shelducks nested before there were Rabbits in the country. They must have used rock cavities along the coastline almost entirely. Has the introduction of the Rabbit, apparently by the Normans nearly a thousand years ago, increased the duck population greatly?

Other ducks are tied by their nesting preferences to the distribution of those birds in whose holes they lay their eggs. The Smew *Mergus albellus* needs the large Black Woodpecker *Dryocopus martius*, and the duck's range in Europe coincides almost exactly with the same well-grown trees and drowned woodlands that the woodpecker prefers. The Bufflehead *Bucephala albeola*, a relative of the Smew and its North American 'ecological replacement', chooses the same type of swampy aspen/lodgepole pine community, and uses the nest holes of another woodpecker, the Flicker *Colaptes auratus*. Two hole-nesters of the sea duck group are currently expanding their range in Britain. Goosanders were first recorded breeding in Scotland in 1871 and had spread to England by 1941; recently they have started nesting in Wales and Ireland. Why did they not arrive before the last century? Why did the UK not have a tree-hole-nesting duck until recently? No-one seems to know the answer. Goldeneyes are also spreading in the Spey Valley in Scotland, but this seems to be due mainly to the provision of artificial nest boxes. The first ducklings were seen in the late 1960s, and by 1981 more than 50 females were hatching about 300 youngsters between them.

It is clear that a shortage of suitable (and that means predator-proof) nest sites can have enormous importance for wildfowl populations. The numbers and types of predator are greater at low latitudes and low altitudes than at the higher ones — possibly one reason why birds migrate out of the tropical and temperate regions in order to breed. For instance, egg-eating reptiles become less common the further one moves from the equator, and arctic and antarctic islands are often free of all but predatory birds, and there may not be many of those. In Australia, 14 out of the 19 wildfowl species are either hole-nesters or marsh-nesters, and these situations will give more protection than ground-nesting. In Britain, however, the majority of breeding species are ground-nesters, and this is even more obvious in northern Europe and Canada. Many of these species are migratory and move to high latitude or, very occasionally, moun-

Above. A female Goosander *Merger merganser* at her nest entrance. She weighs about 1.25 kg. and needs a fairly large hole.

tain breeding grounds where there are very few trees and therefore only a limited number of tree-hole sites. Fortunately, predators are scarce as well.

In an apparent effort at predator protection, some wildfowl nest near birds of prey, or in gull colonies. The best known example of this is the Red-breasted Goose *Branta ruficollis*. It lays in the open and, indeed, the nests may be quite obvious; a small colony of up to ten pairs share their habitat closely with a breeding Peregrine *Falco peregrinus* or Rough-legged Buzzard *Buteo lagopus*. These birds of prey protect the immediate area against nest predators, such as Arctic Foxes *Alopex lagopus*, and hunt for their own food well away from the site. An obvious problem arises when the goslings hatch and the family must reach the nearest water rapidly, without provoking an attack as they leave the close vicinity of their protector's nest. It nests later than other larger geese breeding at the same latitude (Bean Geese *Anser*

Overleaf. Snow Geese *Anser caerulescens* and Whitefronts *A. albifrons* in California in early autumn.

Right. The Bufflehead *B. albeola* female nests almost exclusively in holes made by North American woodpeckers. Here are two pairs.

Below. The female Smew *M. albellus* (this is a captive male), who weighs only 0.55 kg., uses smaller holes than the Goosander, usually old nests of the Black Woodpecker *Drycopus martius*.

fabalis, for instance); in other words, it comes into breeding condition at a longer daylength. Over many years of evolution it has probably selected a later laying date because that gave the birds of prey a chance to establish a nest site first. One consequence of this is its small size; hatching late, young birds have a shorter period in which to grow. The Redbreast is now Europe's rarest goose, and has declined in numbers quite markedly in the last 25 years. Many of the world's birds of prey are likewise showing a slump in population size. Peregrines, in particular, have suffered from the effects of pesticides that result in their laying thin-shelled eggs. Perhaps it is a shortage of Peregrines in the Taimyr Peninsula, where the falcon and the goose nest in such close association, that is causing the decline of the

goose. This Peregrine population winters in parts of the Indian sub-continent where DDT is still an important weapon against diseases such as malaria.

A few ducks, such as the Blue Duck of New Zealand *Hymenolaimus malacorhynchos* which defends a length of stream, are highly territorial and, again, as an anti-predator device, this may be a good thing. Others, however, breed on islands at high density; Gadwall *Anas strepera* may nest as close as 5 m, and occasionally they, and the Tufted Duck *Aythya fuligula*, lay in colonies of gulls or terns. Presumably here the advantage is that the gulls give warning of the approach of a predator and may mob it to drive it away. With the choice of so many clutches, there is also a good chance that the ducks' eggs will not be the ones taken by the thief.

Facing page. The handsome Red-breasted Goose *Branta ruficollis* selects a bird of prey as nesting companion.

Right. Blue Ducks *Hymenolaimus malacorhynchus* are highly territorial along stretches of New Zealand's mountain rivers.

Below. It will be 24 days before this female Tufted Duck *Aythya fulvigula* hatches her eggs.

In general, the importance of concealment and camouflage is paramount for the smaller ducks. The nest is constructed of material lying nearby and so blends with the landscape, but wildfowl usually appear to be incapable of carrying items in their bills. This means that only materials such as grass, sticks, etc. in the immediate vicinity can be pulled in and dropped at the feet, or passed sideways over the wing and shoulder, forming a pile as the bird moves around on the spot. Vegetation growing locally is often shaped over the top of the nest by the sitting bird in an apparent effort to hide herself better from the view of predators searching from above.

In the Magpie Goose, whistling ducks, White-backed Duck, swans and the Cape Barren Goose, the building of the nest is shared by the male and the female. In the true geese, sheldgeese and all the true ducks the task is undertaken by the female alone. This means, inevitably, that goose and duck nests are smaller than those built by swans. A few ducks, such as the stifftails, lay in second-hand nests that coots and moorhens have already constructed among waterside vegetation, and so may seem to have built a very substantial structure.

Down feathers from the female's breast are added as a final lining to the nest with the exception of those 'primitive' species where both parents participate in incubation and thus the eggs are never left uncovered. This feather down, which is unknown in birds other than waterfowl, is pulled over the eggs when the female leaves to feed and so keeps the clutch warm and moist, and again protects it from the sight of predators. The down on the breast becomes especially dark in many ground-nesting females just before laying so that the feathers will help to conceal the nest better. Hole-nesting ducks produce the normal white down; perhaps this also has advantages in that the presence of the light-coloured down makes the clutch easier for the returning bird to locate in the dark of the burrow. In pulling out the down, the bird exposes a patch of bare skin against which the eggs are placed for efficient warming.

Facing page. This European Teal nest *Anas c. crecca* contains no down, so the female has not yet completed her clutch.

Right. A North American Lesser Scaup *Aythya affinis* on her nest. Compare her with a close relative, the Tufted Duck on page 63.

Below. The relatively huge nest of the Mute Swan *Cygnus olor* is built by both male and female. The egg weighs on average 340 g.

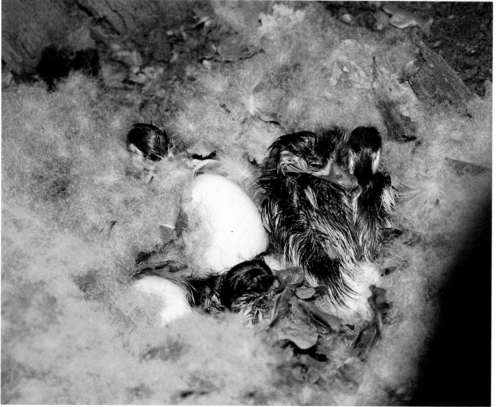

Above. Hatching Tufted Duck eggs *Aythya fuligula*. An egg tooth on the duckling's bill cuts through the shell as the bird revolves inside.

Left. Shelducklings *Tadorna tadorna* hatching among soft white down in a tree hole nest.

Left and above. The North American Trumpeter Swan *Cygnus buccinator* and the marvellous scenery of its nesting habitat.

Egg-laying usually occurs early in the morning and generally one egg per day is laid although with the larger species, such as swans and geese, an egg every two days is the rule. All wildfowl eggs have unspotted shells which are mostly white or pale-coloured. An exception is the White-backed Duck, which lays eggs of a rich warm brown, similar in colour to those of the domestic Maran fowl. This is said to enable the parents to distinguish eggs dumped in their nest by other duck species and, presumably, to push out any that are not their own.

The largest wildfowl eggs are laid by the Trumpeter Swan and weigh 355 g, the equivalent of seven hens' eggs, and among the smallest is that of the Blue-winged Teal *Anas discors* at 26 g. As a proportion of the female's body weight, the largest eggs are produced by the stiff-tailed ducks. The White-headed Duck, for example, lays an egg that, at 94 g, is between 15 and 20 per cent of her own weight, and she typically produces six of them in a clutch.

Wildfowl eggs tend to be ovoid rather than the classic egg shape. Clutch size ranges from the two eggs of many Musk Duck to the 22 laid occasionally by a single Mallard, although the average clutch size of these species is 2.2 and 11.8 respectively. In the Magpie Goose, where one male may be mated to two females that lay

in the same nest, the number of eggs is always about nine, which is presumably as many as can be covered adequately by an incubating bird.

A number of ducks will occasionally lay their eggs in the nests of other birds. In the diving ducks and stifftails this behaviour may be common enough to influence the breeding biology of the entire population. At its extreme the habit is known as 'dump nesting' and is not uncommon if there is a shortage of nest sites and an increasing population, as in Common Shelduck, for instance. A clutch too large to incubate adequately may result. The Black-headed Duck of South America is a true brood parasite, never building a nest of its own but always laying in those of other birds such as ducks, ibises, herons, coots and even hawks. It uses the foster bird only as an incubator and hatcher — no other parental care is required. This is quite unlike the other brood parasites with which we are perhaps more familiar; the Cuckoo *Cuculus canorus* and the Cowbird *Molothrus ater* demand much more of their hosts. Hatching success for the Black-headed Duck is not high, only about 20 per cent of parasitically laid eggs producing duck-lings in one study area: some of the duck's eggs that failed were buried in the nest of the host, and others were laid too late to be incubated fully by the time the host's clutch hatched.

Nests and Eggs

Below. A pair of Blue-winged Teal *Anas discors*.

Facing page: above and below. The eggs of the Canada Goose *Branta canadensis* weigh about 200 g. and are incubated for 28 days by the female alone.

Incubation duties may be shared by the male and the female or performed by the female alone. There are no wildfowl species in which the male alone sits on the eggs; there is one report of a captive cob Black Swan successfully incubating a clutch of eggs without help, but this is not typical. In all 'primitive' waterfowl incubation is a joint effort, while in the white swans, true geese, shelducks and geese and the ducks incubation is a female task. The male still has further choices to make. He can stay and guard the nest, as is the case with all the geese, he can stay nearby and return to the nest at the hatching of his offspring, as in the shelducks and in the Blue Duck and Torrent Ducks, or he can desert the female altogether as does the drake Mallard

and, indeed, most male ducks. The incubation period lasts 22 to 39 days, and its length is generally associated with the size of the bird and with the latitude at which the species breeds. The period is shortest in small wildfowl nesting at high latitudes and longest in large ones nesting near the tropics. It is also longer in hole-nesting ducks than in ground-nesters, perhaps because the pressure to reduce the period of vulnerability is not so great. It has been pointed out already that hole-nesters are more common in tropical regions.

The average duck's egg takes a week longer to incubate than that of the domestic hen, and it is generally much more difficult to hatch successfully after artificial incubation. This is partly due to the fact that the recent evolution of the egg of the domestic fowl has matched that of the commercial incubator very precisely. Any hen in a battery system producing an egg that does *not* hatch successfully in an artificial system will not leave many descendants to carry on her traits. In the wild, the microclimate of the nest can be controlled fairly precisely by the incubating parent bird. The moisture content is influenced by her swimming activities, and both the humidity and the temperature of the eggs by the closeness with which she sits, how often she stands to turn them, and by how often she leaves for a meal and a drink. In nests of swans and geese, research has shown that the clutch is raised very slowly to incubation temperature during the first two days, and that there is a considerable difference in warmth between the bottom and top of the egg. There is a rhythm in temperature and humidity which is related to day and night, and the nest material tends to dry out during the course of incubation. Finally the egg, which is turned just less than once an hour, is rotated mainly around its long axis rather than its short one. None of these features is found in the typical commercial incubator.

For about a month the female wildfowl has been sitting on a delicious meal for some predator. Things do go wrong, and a certain proportion of the eggs will fail to hatch for some reason or other. The female may give up her task if her body's food reserves reach exhaustion — because of a prolonged drop in temperature, for instance. The beautifully camouflaged Eider Duck *Somateria mollissima* sits on her nest for 25 to 28 days and does not eat at all, losing one-third of her weight in that time. She barely moves, until a predator approaches very closely, since movement destroys the effect of the camouflage, but, if a potential thief does make her fly, she may defecate over the eggs as she springs away. The gut contents of a starving duck have an odour that is quite appalling and, while the droppings are wet, the smell may deter any mammal, such as a dog or an Arctic Fox. Hole-nesters have another line of defence — a startling snake-like hiss from the darkness of the hole.

Right. The well-camouflaged Eider *Somateria mollissima* incubates without feeding for at least 25 days.

Overleaf. A captive pair of Ruddy-headed Geese *Chloëphaga rubidiceps*. In Tierra del Fuego, their eggs and nests are threatened by an introduced fox.

Introduced predators have brought island wildfowl to the verge of extinction with depressing frequency. Most of these ducks and geese were ground-nesters, and their eggs made easy meals for the newcomers. Foxes have nearly eliminated the Aleutian Canada Goose *Branta canadensis leucopareia* which once bred on all the outer Aleutian Islands. In the early 1900s, Arctic and Red Foxes *Vulpes vulpes* were introduced to all except the tiny island of Buldir for the purpose of commercial fur trapping. By 1963, goose numbers were as low as 300 and all of them were being hatched on Buldir. A programme of fox-eradication, captive propagation and protection during the non-breeding season in California is leading to a slow recovery in the population. The Ruddy-headed Goose *Chloëphaga rubidiceps* on the island of Tierra del Fuego is threatened currently by another fox introduction.

Turning to the Caribbean, the Indian Mongoose *Herpestes sp.* was released on Jamaica in 1872, from where it spread to many Caribbean islands. The Cuban Whistling Duck *Dendrocygna arborea* is among the many ground-nesting birds that as a consequence are listed in the Red Data Book of endangered animals. The same predator was taken to Hawaii in 1833 to control the introduced rats, but it turned to the easier diet of eggs — including those of the Hawaiian Goose or Nene, and the Hawaiian Duck or Koloa *Anas platyrhynchos wyvilliana*.

Black Rats *Rattus rattus* restrict the Campbell Island Flightless Teal *Anas aucklandica nesiotis*, one of the world's rarest birds with an estimated 30–50 individuals, to a tiny, inaccessible stack of rock. Cats have 'pushed' its relative the Auckland Islands Flightless Teal *A. a. aucklandica* onto the offshore islands of that archipelago. The Auckland Islands Merganser *Mergus australis* became extinct as recently as 1902; the foreign cats, dogs and pigs proved too great a hazard.

Facing page. Male and female Hawaiian Duck or Koloa *Anas p. wyvilliana* at Martin Mere. It is thought that the race has evolved from the Mallard, although the drake (on the right) has lost much of his bright plumage.

Left. A pair of Auckland Islands Flightless Teal *Anas a. aucklandica* (the male is in front) among the kelp. They are loafing, not feeding, and will take to the water should a predatory Skua fly close.

CHAPTER 4

Parental Care

From the time that the first sounds are heard within the eggs, hatching may take two days. The young ducklings, cygnets or goslings are then brooded, their down dries and they sleep. The parent that is sitting on the eggs calls to them as soon as the process of hatching starts, and the whole family becomes familiar with one another's voices. Incubation does not begin until the entire clutch is laid so that, as with most birds with active young, hatching occurs at the same time in all eggs and may be influenced in its precise timing by 'clicks'. These clicks are made in the throats of the embryos as soon as the tip of the bill breaks into the air space at the blunt end of the egg. By clicking at different rates, ducklings can communicate with one another

and hatching may be slowed or hastened by a few hours. This mechanism enables the brood to leave the nest together — usually in the early morning — and at almost exactly the same age.

How do ducklings of tree-hole-nesting species reach the ground? Stories of the young being carried in their parents' bills or riding down on their backs are not uncommon, but they do seem most unlikely. The simple answer is that the young are normally coaxed to jump by calls from the ground beneath. Day-old Carolinas and Goldeneyes have absolutely no fear of heights and leap from the nest entrance readily. The little Mallard is much more dubious about the business of falling and, should its mother have nested off the ground in the crown of a pollarded willow, for instance, it teeters about on the edge for quite a while before she can talk it down.

The ducklings have already become familiar with her voice during the two days of hatching and, although they may never have seen her very clearly, it is obvious that 'imprinting' has occurred. The word is translated

Right. A Brent Goose *Branta bernicla* standing guard over one gosling and more hatching eggs.

Below. A flock of Maned Geese or Australian Wood Duck *Chenonetta jubata*. The female (in the centre) has a long call that sounds very like a cat.

Parental Care

The shallow sea off the Falkland Islands gives sanctuary to a family of Greater Magellan Geese *Chloëphaga picta*. The gander is the white bird nearest to the camera.

from a German one that means the stamping of a figure on a coin. The duckling has, in a very short time, become 'stamped' with an image of its siblings and of its parent and, thereby, is made aware of its own identity.

The leading of offspring from the nest to water and food is a crucial parental role and all young wildfowl (except Black-headed ducklings) follow their parents; typically the mother leads, with the father, if he is present, behind. 'Imprinting' has great importance here. The earliest studies of this special form of learning were made on ducklings and goslings by the great Austrian ethologist Konrad Lorenz. He found that, although the true parent may bring about the 'best' following response, none of these little birds seems to hatch with a real recognition of the object to be followed, and domestic foster hens and human beings are acceptable as substitutes. It is interesting that in sheldgeese like the Magellan Goose *Chloëphaga picta* and the Kelp Goose, where the male and female are quite different in plumage, leg colour and voice, the goslings seem to have no difficulty in learning to recognise both their parents. They normally follow their mother (who was presumably the first object that they saw and heard as they hatched) but, if she is removed, they tag after the male equally closely and snuggle under him for warmth if he is sitting down.

Voice as well as vision are obviously important in parental recognition — young shelducks leaving a dark burrow have certainly never *seen* their mother, they merely move after her calls until they reach the light. They then have their first sight of her, and of their father who will join the family and help lead his offspring to the sea. Although sounds are clearly important, most research has investigated what the young bird can see and what it prefers to follow using its eyes. The colour, shape, size and movement of substitute parents have all been tested for the response they obtain. Having followed an object for only a very short time, certainly for less than two days, the ducklings will not imprint on anything new but react as if they are frightened. They will follow only their first preference — even if that is something as incongruous as the researcher's red watering-can.

Parental Care

The calls of the young themselves are also important in keeping the family together. Contact calls are heard whenever the group is feeding or searching for food. If a youngster is lost, it gives a particularly loud, sharp distress call and the parent will go and find it; only the Black-headed duckling, which receives no parental care after hatching, hardly ever calls in distress. Young swans, true geese and the Cape Barren Goose are unusual in seeming to coordinate their resting periods by the use of long, trilling 'sleepy calls'. These low calls are given when the cygnets or goslings nestle together, with half-shut eyes and bill-tips tucked in one another's down. Sometimes sleepy calls are heard from a bird that is actively feeding or swimming about, and it can then be predicted that the youngster is tired and will soon be on its parent's back or sitting ashore with its mother brooding it.

Parental care in the Magpie Goose, whistling ducks, all the true geese, swans, sheldgeese and shelducks, and a few tropical or low-latitude ducks, is shared by

the male and the female. In most of these species the male and female look alike, although there are rare exceptions — the Magellan Goose and the Kelp Goose have already been mentioned. The tiny Ringed Teal of South America is another example of a species where male and female are permanently different in plumage, the drake being the more conspicuous yet also fierce in defence of his brood.

Where there is dual parental care, it is the larger male that undertakes most of the protection of the young, and this can be quite furious in the case of a large male swan or goose. But I have also seen a drake Chiloe Wigeon *Anas sibilatrix* fly at the face of a person that had picked up one of his ducklings. Probably this behaviour is not usual, but the bird had been hatched in captivity and was presumably not especially afraid of humans. The normal response to such a raid would be to perform a 'broken-wing display' a short distance from the young in order to lure the danger towards an apparently injured adult who, at the last minute, is found to fly perfectly. Meanwhile, the rest of the family will have reached the water or the safety of long vegetation.

Above. A captive male Chiloe Wigeon *Anas sibilatrix*. The sexes of this grazing duck look alike and both care for the young.

Facing page. Male and female Kelp Goose *Chloëphaga hybrida* are strikingly dissimilar. The male is white and very territorial while the female, who sits on the nest, is relatively well-camouflaged.

Above. Mother's wings provide shelter for a brood of Mallard ducklings *Anas platyrhynchos*.

Opposite. A female White-headed Duck *Oxyura leucocephala* (the male is on page 41) defending her young.

Not that the water is always safe. Pike and large eels may lurk beneath the surface and pull down small birds, or snap them up as they dive. Turtles and crocodiles take their toll in some countries. One or two species of wildfowl solve this problem by carrying their young on their backs. Black-necked Swans, whose cygnets grow slowly and are vulnerable for a long time to underwater dangers, especially at night, carry their young until they are about six weeks old. Mute Swans and Black Swans do so as well, although not so often nor for as long. At least two ducks carry their young regularly, the Musk Duck and Salvadori's Duck *Anas waigiuensis* of New Guinea. Since the male Musk Duck assumes no role after copulation, it is the female alone that carries

the young. Her small brood rides around on her back, even diving with her by holding on to her neck feathers with their bills. In Salvadori's Duck also it is the female that carries and again the brood contains very few ducklings. The female Goosander, however, sometimes takes almost the whole of her large family aboard, apparently to remove them rapidly from danger rather than to warm them.

The water is also liable to be cold, of course, and the young of 'carrying' swans have the additional advantage of sleeping warm and dry among the feathers of their parents. The young of species that do not carry must come ashore to be brooded. A Mallard duckling will need additional warmth at night in order to survive the first couple of weeks of life. This is true of most young wildfowl with the exception of the very aquatic diving species, such as the stifftails, whose offspring are so well covered in fat as an insulation against cold water that they need little extra heat. There are instances of deserted young Ruddy Ducks rearing themselves and, of course, those unusual stifftails, the parasitic Black-headed Ducks, do so all the time. Brooding is almost always a female's task, but male Magpie Geese and male whistling ducks become closely involved. Female wildfowl, as has been pointed out already, are almost always brown, camouflaged and inconspicuous, and this is presumably in order to escape detection by predators liable to attack either her or her young.

The young themselves are also usually drab brown or patterned in such a way that they are not eye-catching. In general, the further north a species breeds the darker is the ducklings' down; the Eider in the photograph on this page is a good example. Perhaps this phenomenon is related to the heat-absorbing properties of brown and black. Eyes tend to be obscured by dark eye stripes, as in Mallard ducklings, or by a totally sooty crown to the head as in the young Golden-eye. Tummies are usually pale, perhaps so that under-water predators can detect less readily a body floating above them. The down of those few wildfowl that hatch at high altitudes (Bar-headed goslings *Anser indicus*, Andean goslings, and Andean Crested ducklings *Anas specularoides alticola*) is especially long and silky, perhaps under the same environmental pressure that makes the wool of the Mountain Alpaca *Lama pacos* and the cash-mere goat so soft. The downy coat of the young Hawaiian Goose or Nene can occur in two forms: normally the gosling has a dense covering of close-packed down but a few hatch with down that is thin. This condition, which does not affect the adult, seems to result from a recessive gene that causes some of the short plumules to be missing so that the longer ones are more prominent. The down looks scruffy, and probably is not waterproof. It is possible that thin down was no disadvantage to the original Hawaiian Goose population that descended to lower altitudes to breed; the sea level climate at 20° North is warm enough for affected young of this non-swimming goose to survive. Indeed, it is quite likely that thin down was useful in hot weather. However, one wonders whether the condition is not now implicated in the endangered Nene's lack of success in the wild: habitat destruction and the presence of the Mongoose

Left. A captive Andean Goose *Chloëphaga melanoptera* with goslings.

Above. An Eider duckling *Somateria mollissima*.

Parental Care

Preceding page. Barnacle Geese *Branta leucopsis* on the Island of Islay in winter.

Below. Waterbirds in Africa: Fulvous Whistling Ducks *Dendrocygna bicolor*, White-faced Whistling Ducks *D. viduata*, a Garganey *Anas querquedula* (right foreground) and Pink-backed Pelican *Pelecanus rufescens*.

have forced the few remaining birds to breed at colder, wetter altitudes where thin down is probably a further problem.

Duckling size at hatching is obviously dependent on the size of the egg. Eggs of island ducks and geese are always larger relative to their body size than those of the continental ancestors from which these species were derived. David Lack suggested that this is related to the smaller clutches that island wildfowl lay, and that the larger egg will be advantageous because the young are bigger at hatching and so survive better if feeding conditions are less favourable. He instanced the average size of newly-hatched Laysan Teal *Anas laysanensis* of 30 g. in comparison with young Mallards

which weigh 34 g, although, as an adult, the Teal of Laysan Island is only *half* the weight of a mature Mallard, the species from which it supposedly evolved. Species hatching in colder places also have heavier cygnets, goslings or ducklings than their relatives hatching nearer the equator. The larger youngster conserves heat better, and probably hatches with extra food stores in its liver and beneath the skin that will last it for a few days while it learns to feed.

The business of successfully learning to find food will be considered in more detail in the following chapter. The young bird's rate of growth is dependent not only on the food supply, but also on the hours of daylight available for feeding. Ducks, geese and swans hatching in Iceland or Greenland, for instance, can feed for longer than those hatching in Scotland. The fledging period (the length of time that it takes a young bird to reach the flying stage from hatching) tends to be much shorter in species breeding at high latitudes. The Hawaiian Goose living at 20° North in the tropics fledges in about 12 weeks, the Barnacle Goose at 75° North does so in half that time, although the final weight of both adults is much the same. Furthermore, the nutritional requirements of birds growing at different rates are not identical, and the grass in Greenland, where the Barnacle gosling is reared, has a higher protein content during the bird's adolescence than the herbage of Hawaii.

The family will stay together during the whole of the fledging period. In those ducks where only the female cares for the young, it is normal for her to leave just before or as soon as they can fly. In swans and geese, the family will remain as a group for about nine months — the northern swans and geese fly south together to the wintering grounds where the parents, especially the male, will continue to defend their young in the winter flocks. They will go north in company in the spring and separate only just before the nesting territory is reached.

It is generally assumed that parental care in wildfowl never involves the feeding of the young. This behaviour is, however, well established in the Magpie Goose and the Musk Duck, and exists in a simple form in the swans. In Magpie Geese, bill-to-bill feeding occurs for about six weeks. The gosling has a shrill begging cry, and even newly-hatched birds lift their bills, gape and call, which seems to stimulate the two or three adults in the breeding group to offer food. Items are not carried to the young, which come to the parents' side to be fed. The food offered does not differ much from that which

the adults take themselves — goslings pick up items from the water surface and from low-growing plants, while the mature birds strip seedheads beyond the reach of the young and bring up material from beneath the water. Magpie goslings are unusual in another way: they are very aggressive towards one another, especially when more than one is begging from a single adult. A brood is quite difficult to hand-rear as they peck and squabble a great deal. This fighting is almost certainly related to the evolution of parental feeding and much the same behaviour occurs, for instance, in

Facing page. Brent Geese *Branta bernicla* and duck in winter grazing tidal vegetation.

Above. Magpie Geese *Anseranas semipalmata* are migratory according to the rainy seasons in northern Australia.

Parental Care

Oystercatchers *Haematopus ostralegus* if more than one chick begs from the same parent.

The Musk Duck female cares for her brood for about eight weeks and also feeds it strenuously during this time. She dives and rises with food that the youngsters plunge at and grab. The brood, which may start as two or even three ducklings, is frequently reduced to a single juvenile so perhaps fighting develops here too, one youngster receiving most of the food and so surviving at the expense of its siblings.

Young swans also receive parental assistance in finding food, but the feeding here is of a different type. Oskar Heinroth reported as long ago as 1911 that most breeding swans pulled up water plants for their cygnets. They also pull vegetation down from the bank using much the same sideways-passing movement seen at nest-building. Another feeding behaviour that increases when a swan has a brood is foot trampling. This trampling creates an eddy that raises invertebrates and debris to the surface where any food items are snapped up eagerly by the young swans. Trampling sometimes occurs when the family is ashore and being fed by hand in a situation that appears quite inappropriate. Heinroth called it *Locktrampeln* or 'attraction trampling' — the movement may have become 'ritualised' during evolution so that it now signals the young to come close.

Why is parental feeding so uncommon in ducks, geese and swans? In the previous chapter the problem that wildfowl have in carrying objects in their bills was mentioned in relation to nest-building. In the same way, apparently, food cannot be carried to the young and the young must always go to the adult. Even the female Red-crested Pochard must go to her mate to receive his tribute of weed. Perhaps, therefore, it is not surprising that neither courtship nor parental feeding has developed further.

Below. The Whooper Swan *Cygnus cygnus* can raise its upper mandible while feeding on aquatic vegetation.

Facing page. Female and young Harlequin Duck *Histrionicus histrionicus* negotiating white water on the river Laxá in Iceland.

One of the most important aspects of parental care is the protection of the brood from predation. Almost all wildfowl utter alarm notes in the presence of danger and this makes the young scatter, run for cover, or freeze. Sometimes alarm causes the ducklings to gather into a tight bunch. Almost all young wildfowl are gregarious and show a pronounced tendency to associate with others of the same age. 'Crèching' is the pooling of several broods and their apparent care by only a few adults. The British population of Common Shelduck undergoes an annual migration in order to moult in Heligoland, off the German coast. This migration involves almost all the breeding and non-breeding adults. The ducklings, many still in down, are deserted by both parents and usually join to form large groups with which one or two adults remain. Although not a colonial nester, Shelducks meet their feeding requirements in a narrow shoreland habitat where a number of families collect, and the amalgamation of broods occurs quite naturally. Another duck with crèching young is the colonial nesting Eider. Its 'downies' are relatively inde-

be taught to feed; even ducklings that use quite sophisticated techniques to catch live fish learn to do so without instruction. At hatching their bills are more alike than they will be as mature birds. The 'basic wildfowl' has a longish neck and a bill that is rather blunt and spatulate with a rounded tip and a 'nail' that often gives the whole thing a slightly hooked appearance. This bill is provided with fine comb-like structures, called lamellae, that lie in parallel rows along the upper and lower edges and which, in adults, are developed in a variety of ways. Wildfowl bills never act as fine tweezers, crushers, stabbers or even as deep probers in mud; but they can do much else besides, and the Mallard, with its rather generalised bill (its specific name of *platyrhynchos* means 'flat beak'), is one of the world's most adaptable and successful birds.

Right. The omnivorous Egyptian Goose *Alopochen aegyptiacus* is a successful African species.

Below. The beautiful drake Cinnamon Teal *Anas cyanoptera* is a dabbling duck related to the shovelers.

Feeding

The young wildfowl hatches with certain preferences that are more or less innate. It will peck at a variety of objects that contrast with the background, and some objects will receive more pecks than others. Green is the colour that a cygnet, gosling or duckling will peck at most; if the object is small, and has in addition a worm-like shape, then it is preferred. Something that moves will often be pecked at particularly vigorously by duck-lings whatever its colour, so wiggling red worms are very attractive to them although not to goslings, to whom movement is not so important. In goslings, the innate preference for green seems to direct their attention to blades of grass. Taste we know little about, but flavour and texture also are probably important in deciding what the young bird will swallow. Smell appears to be little used (as in most birds), but research results are inconclusive.

Temperate-zone dabbling ducks need less protein than growing chicks of the domestic hen, and are good converters of food. Much of their initial energy intake goes towards an insulating layer of fat on the tummy — at and below the swimline — which protects them from the cold water on which they sit for much of the time. Divers have more need of an overall fatty layer to insulate them from low temperatures beneath the surface — and diving ducks are well-known to the wildfowler

as succulent 'butterballs'. The rigours of migration also require the prior storage of plenty of energy. The breeding season and period of wing moult bring special needs for protein, energy and minerals. In many females, the calcium for eggshell formation will come from her own bones and be replaced later from the mineral content of her food. The Pink-footed Goose, returning to a bleak Iceland nesting site that is barely emerging from the snow, will eat the eggshells left in the nest from the previous year's hatch. Grit to aid digestion and the breakdown of food in the gizzard is vital, and a shortage of suitable grit may limit the exploitation of any habitat, especially by seed-eating and grazing birds.

Above. The only diving duck of New Zealand, the New Zealand Scaup *Aythya novae-zeelandiae*, feeds in clear water.

Facing page. The male Meller's Duck *Anas melleri* from Madagascar is a rare dabbler with a longer bill than its relative, the Mallard.

Diving

Fresh water divers include most of the pochards and scaup, a few of the whistling ducks, the stifftails, the White-backed Duck and the ducks of fast-flowing rivers. The first three take underwater roots, tubers and rhizomes but also some molluscs and other invertebrates (the recent spread of the Tufted Duck in Britain is said to be linked to the accidental introduction in 1824 of the foreign Zebra Mussel *Dreissenia polymorpha* to the London docks). Feeding at night by touch or by taste is typical of this group.

Chironomid larvae (the young stages of gnats, often called 'blood worms' by fishermen), which live in the mud at the bottom of many ponds and rivers, especially those mildly polluted by farm or sewage effluent, are a favoured food of many inland divers such as the Ruddy Duck and White-backed Duck. Presumably the ducks' droppings also are part of the cycle of gnat production.

The group of ducks found in river torrents is of particular interest, partly because they are all highly territorial, but also because of their exploitation of a habitat that elsewhere in the world is left to insectivorous fish and certain amphibians. The Blue Duck of New Zealand, Salvadori's Duck of New Guinea, and the Torrent Ducks of South America live in water devoid of such animals as trout. They consume the caddis-fly, stonefly and mayfly larvae that would in other rivers be fish food. The existence of these ducks is threatened in some places by the introduction of game fish for sporting interests. The only northern hemisphere duck found in 'white water' is the Harlequin *Histrionicus histrionicus* of North America, eastern USSR, Greenland and Iceland. It feeds in fast-flowing rivers only in the breeding season, taking the larvae and pupae of the black fly; at sea during the winter it feeds on shellfish.

By no means all the sea duck divers catch sea fish, but the group is so closely associated with marine conditions that it is convenient to consider them together.

In the fish-eating mergansers the lamellae of the basic wildfowl bill have been converted into rows of saw-edged clamps which are excellent for holding slippery fish firmly. In addition, there is usually a hooked nail against which the fish can be lodged before it is swallowed head first.

Many of the true sea ducks (eiders, Harlequin, goldeneyes, Longtails and scoters) are shellfish eaters. Unlike Oystercatchers, they do not remove the succulent mollusc or crustacean from its shell before eating, but consume it whole. And unlike some other seabirds such as gulls, they do not cast a pellet of the shells but pass the lot straight through the gut, squeezing out and digesting the flesh as the meal goes down. They need no extra grit to break the shells, the muscular gizzard grinds sufficiently powerfully to crush the invertebrates against one another.

An ability to drink salt water is vital to seabirds; they excrete the surplus salt, not via their kidneys as mammals would, but through special glands situated beneath the skin above their eyes. These glands were inherited by birds from their reptilian ancestors. Their presence means that all wildfowl can adapt (if slowly) to a brackish water supply. The glands grow as their workload increases, and it is often apparent that a flock of swans, for instance, has been feeding at sea because their foreheads have a distinctly knobbly profile. It is said that that isolated mountain sheldgoose, the Abyssinian Blue-winged Goose, can be poisoned by salt because, after millions of years of not needing the capacity to deal with a high salt intake, it has lost the use of its excreting glands. I have never put this to the test, nor do I know of anyone who has done so, but the statement sounds likely to be true.

Almost all diving ducks have pale abdomens. This white patch includes the area where there is a layer of fat beneath the skin that keeps the bird warm. The white belly may be useful in that fish cannot see a merganser coming. Or perhaps the countershading makes it less easy for underwater predators to spot a possible meal. The goldeneyes have large air spaces behind the nostrils which give the birds a rather pronounced hump on the front of the head. These may provide the bird with a supplementary air supply when it dives. The goldeneyes, however, dive only in rather shallow water; the Long-tailed Duck manages without special sinuses to go as deep as 20 m and stay down for as long as 60 seconds.

Facing page. Two male and a female Harlequin Duck *Histrionicus histrionicus* in captivity.

Above. The female Red-breasted Merganser *Mergus serrator* has a saw-edged bill for holding fish.

Below. The Tufted Duck *Aythya fulvigula* has spread in Britain since Zebra Mussels were introduced.

Dabbling

Among the dabbling ducks, the most common dietary preference is for seeds which, as plant food goes, are relatively high in protein and energy content. Seeds are seasonal in production, at least in the temperate zones, and diets may need to change in the spring as the supply wanes. The finer the seed that is filtered, the finer the lamellae on the bill. The Mallard and Spot-billed Ducks *Anas poecilorhyncha* of India take a fairly large seed. On the other hand, the Green-winged Teal *A. crecca* filters smaller food items than most, and relies on the tiny seeds of wild rice, sedges and pondweeds, while also eating midge larvae and small snails. Shovelers have even finer lamellae and typically feed while swimming with the bill half in and half out of the water. They are sieving the surface film for food particles and often swim in pairs or small parties because the movements of the other birds concentrate any floating debris into 'wake' lines.

The Australian Pink-eared Duck *Malacorhynchus membranaceus* is another dabbling duck with a large spatulate bill which superficially resembles a shoveler's. In fact the birds are probably not very closely related; their bills are evidence of convergent evolution and are adapted to feeding on similar items in a similar manner.

The shelduck group are mainly dabblers that are capable of some grazing. Between 80 and 90 per cent of the food of the Common Shelduck in Britain is a small salt water snail known as *Hydrobia*. This mollusc lives on the surface of the mud at high densities and Shelducks filter them out while sweeping their bills from side to side at foot level. When the weather is very cold and the seashore freezes, the ducks suffer starvation and large numbers may die.

Almost all dabbling ducks, whether or not they are vegetarian as adults, are insectivorous until they are about three weeks old. This is presumably because plant protein is not so easily metabolised by a growing duckling, and perhaps because of a general shortage of seed material in the northern spring. The insect 'flush' in the late arctic spring is irritatingly familiar to anyone who has been there. Many migratory birds, not just the ducks, hatch their eggs to coincide with the time when insect larvae change into the adult stage. The swarms seem endless, and duckling production at places like Mývatn (which means 'midge lake') in Iceland can be staggering.

Facing page. A pair of feeding Shovelers *Anas clypeata* and a female on her nest.

Left. An Indian Spotbill *Anas poecilorhyncha* filter-feeding.

Below. A feeding flock of male and female American Green-winged Teal *Anas crecca carolinensis*.

Feeding

Grazing

Grazing wildfowl have the longest necks of all and have developed their lamellae to perform a shearing action. The short-billed grazers, such as the Red-breasted Goose and Ross's Goose *Anser rossi*, are common and successful on short grass, while the longer-billed birds are found on longer herbage. Brent Geese *Branta bernicla* graze sea grass *Zostera* in winter and their daily feeding rhythm is affected by the tides. Not all grazing wildfowl are geese. The wigeons are three species of dabbling duck whose short bills are designed for grazing although, unlike geese, their young are insectivorous. The Maned Goose, or Maned Wood Duck, of Australia is, as its second name suggests, a perching duck, but it also has a goose-like bill and a grazing habit. Again the duckling is not reared on the proteins of pasture grass but on aquatic invertebrates.

The temperate grasslands of the southern and the northern hemispheres have grazing geese that are not closely related. The 'true' geese seem, as I have said, to be a fairly recently evolved group that eliminated the sheldgeese from the north many thousands of years ago, presumably by competing more successfully. The southern hemisphere Cape Barren Goose almost certainly represents the earliest development of the grazing habit in wildfowl. Like the Blue-winged Goose, it may once have had a much wider distribution. Fossils of birds that were related to the Cape Barren Goose are found in New Zealand and Hawaii, but the species is now confined to the sea-washed turf of a few offshore islands in southern Australia. The grazing activities of geese in general have been very greatly expanded by the increased provision of fertilised grassland by agricultural Man. And many geese are now turning to other farm crops — young cereals, roots and grain — and coming into conflict with the rural economy.

Compared with the grazing mammals, geese seem inefficient consumers of grass. Their food is just squeezed dry and no cellulose digestion takes place, but 'true' geese, sheldgeese and Cape Barren Geese are unique in raising their young on vegetable protein and this capacity has enabled them to be a most successful group in the temperate regions of the world. As grazers they spend much of the day feeding; seed-eaters, on the other hand, often fill their crops only twice daily and sleep for much of the rest of the time.

Roots are an important food for many grazing species, in particular for Snow Geese *Anser caerulescens* and Magpie Geese. The heavy, thickened bill margin of the Snow Geese is useful for digging out and cutting up the roots of the club-rush *Scirpus* from marshy places, while tubers of the spike-rush *Eleocharis* are the single most important item in the diet of the Magpie Goose. The beak is massive, hooked and well adapted for root eating, as can be seen in the photograph on page 13.

Right. The Bewick's Swan *Cygnus c. bewickii* has a grazing bill capable of dealing with longish vegetation.

Below. Cape Teal *Anas capensis*, in east Africa, are seed-eaters.

Below right. The Brent Goose *Branta bernicla* is a tidal zone grazer.

Feeding

Facing page. An autumn gathering of ducks, mainly Pintail *Anas acuta* at Lower Klamath, northern California, suggests an abundant source of seeds or grain.

Left. The heavy bill of the Snow Goose *Anser caerulescens* is for root-eating.

How do we *know* what wildfowl eat? Stomach contents are a useful indication and, because many ducks and geese are shot for sport, their guts are often available in some numbers, and the last meal can be sorted and its constituent parts identified. Some items are digested faster than others. Seeds may remain for hours in the crop, while chironomid larvae are attacked by digestive juices much more rapidly. Droppings can also provide clues to the origin of the bird's recent food supply. For instance, the grasses that geese have consumed can be distinguished by microscopic examination of the leaf cells in the droppings. Or insect remains can be found in the 'splashes' left behind by Harlequins or Blue Ducks on their loafing rocks in the middle of a river.

Do flocking species of wildfowl gain any advantage by feeding in such large concentrations? The areas that are suitable for feeding may be scattered over a wide area and young birds benefit by following others to good safe sites. An Eider Duck looking for mussels would do well to join a group already feeding successfully. Studies of flocked Barnacle Geese have shown that the frequency with which the bird looks around is associated with a general level of feeding success. In a large flock, the individual bird needs to search for danger much less than in a small one. By placing model geese in fields where geese regularly feed, it has been shown that birds flying over are more likely to join models that have their heads fixed in the feeding position than models that are 'looking around'. Sea ducks often synchronise their dives when feeding on live fish and invertebrates since their prey is then less likely to escape; Steller's Eider *Polysticta stelleri* does this in a quite spectacular way. In some cases, however, the presence of very large numbers of wildfowl is merely an indication of abundant food, after a grain or seed harvest. The largest wildfowl flocks, as in the splendid photograph on page 107, are seen in early autumn during the migration season.

CHAPTER 6

Flight and Migration

The ability of birds to fly has been the major influence on their evolution, and the habit of migration increases their ecological flexibility enormously. They are able, for instance, to exploit the resources of the northern tundra (where there are long hours of daylight in which to feed, less competition for living space, a relative lack of predators and plenty of protein-rich food sources) and then to move to an equally warm or even warmer climate in which to spend the non-breeding season. The longest duck migration is that of the Blue-winged Teal which nests up to 60° North in Canada and the United States and then finds a second summer beyond 30° South in South America. Here it is common in Peru and occurs as far south as Chile and Argentina. An adult drake, ringed at Oak City, Manitoba, was recovered in Lima, Peru, 6,500 km (4,000 miles) away. The Garganey *Anas querquedula*, its Old World relative, is astonishingly widespread in the non-breeding season, occurring south of the Sahara in Africa and straggling to many oceanic islands, such as the Seychelles, Hawaii and Barbados, apparently miles from its normal range. Interestingly, neither the Garganey nor the Blue-winged Teal seems to have been the ancestor of any of the eight tiny populations of island dabbling ducks, some of which are shown on page 74, perhaps because they are so highly migratory that they always leave again and never settle down to breed. There is a possible exception on the islands of Amsterdam and St Paul in the sub-antarctic ocean, where sub-fossil bones found recently may belong to a small duck similar to the Garganey.

Island ducks were of special interest to David Lack, who was intrigued by the fact that on most remote islands there is only a single resident species. This is surprising since many other ducks reach these islands in their wide and frequent wanderings, facilitated by their powerful flight and ability to settle on the sea. Resident dabbling ducks of islands are all of medium body size, irrespective of the size of their original mainland ancestor, and usually have a generalised diet. The males have tended to lose their bright plumage and resemble females and juveniles. Lack suggested that

Egyptian Geese *Alopochen aegyptiacus* at sunset. The species once occurred along the entire Nile but is now uncommon there except in upper Egypt.

new arrivals are unable to establish themselves because, on an island, one duck species filling a broad niche excludes any specialists arriving subsequently. The Campbell and Auckland Islands Teals have developed flightlessness and almost all island wildfowl have reduced their wing power to some extent. Compared with other black geese, the Hawaiian Goose, for instance, shows a reduction of about 16 per cent in the bones and muscles of its wings. Three of the coastal-living steamer ducks of South America have likewise lost their ability to fly. Why? It seems that there must be some positive advantage in not bothering to fly — in not developing beyond the fledgling stage but retaining the juvenile condition of flightlessness. Nutritionally, the time of fledging is very hazardous since large amounts of energy are needed to form the huge flight muscles. Why bother to develop these muscles if there are very few predators to fly away from, and there is no need to migrate? In addition, for a flying bird, there is always the danger of being blown from a small island or coastal territory in

high winds and being unable to find one's way back. But the hazards of flightlessness as a way of life have also become quite clear. More flightless birds, including the Dodo *Raphus cuculatus*, have become extinct in historic times than any other category. In the recent hostilities in the Falkland Islands, the Flightless Steamer Duck *Tachyeres brachypterus* is said to have been used for target practice by both sides.

All adult waterfowl, except the Magpie Goose and a few sheldgeese, have a short annual period of flightlessness, usually following breeding, when they lose their primary and secondary wing feathers simultaneously. Before this happens they put on weight and may change their diets in order to include the necessary nutrients to replace old feathers with new ones grown during only two or three weeks. They may move onto water for safety or migrate to other areas that offer security from predators. The late summer moult migrations of the European Shelduck populations to the Heligoland Bight are particularly well known. Shelducks have been found

to leave England on fine evenings with little or no cloud, good visibility and a following wind: the wind direction apparently is critical. The breeding birds, as was pointed out in Chapter 4, may even desert their young before they are fledged and, during August and September, up to 100,000 Shelducks may gather in this relatively small area. Is there any reason, apart from safety, why birds should move long distances in order to moult? It is probable that an enhanced food supply which is available relatively late in the summer is another of the benefits to be gained.

Why migrate at all? It is clear that the advantages must outweigh the costs in both summer and winter during the non-breeding period. To high-arctic-breeding geese the benefits do not always seem obvious. Migrating so far north that the weather is unreliable means that they cannot breed successfully in every year; indeed they are successful in raising a family in only half their summers, on average. Nevertheless, in those 'good' years, excellent food sources and the lack of predators

Preceding page. This small duck, the Garganey *Anas querquedula*, is one of the great travellers.

Facing page. These Flightless Steamer Ducks *Tachyeres brachypterus* are in wing-moult. The female, on the right, is calling, perhaps to her half-grown young.

Below. Emperor Goose *Anser canagicus* incubating in Alaska. In winter she will move to the coast of the Aleutian Islands and the Alaska peninsula.

mean that sufficient goslings are fledged to keep the populations stable in the long term. In the winter, again, they travel as far as is necessary to find food, security and a reasonable climate. The small-bodied geese tend to fly further south than the larger ones since their size makes them more vulnerable to the cold. In Britain, the Whooper Swan can winter further north than the Bewick's for the same reason of greater body mass.

Wildfowl vary in whether they migrate as a family or not. In species with a prolonged family life, such as swans and geese, tradition seems important in the selection of the wintering grounds. In ducks, where migration *en famille* is not common, the sexes move different distances because the sexes compete for food and perhaps because the female's smaller size requires a warmer climate. Drake Teal and Pochard *Aythya ferina*, for instance, have a more northerly winter distribution in Europe than their ducks. A few individuals of a migratory species may stay near their breeding sites in areas of thermal activity, such as is the case with some Whooper Swans in Iceland and Trumpeter Swans in Yellowstone National Park. Power station effluent is a modern equivalent of hot springs, and swans and ducks accumulate around outflows in winter because the water is warm. A freeze-up means that mammalian predators can walk on water and become a serious threat. The sea rarely freezes, so many temperate-zone wildfowl move to the coast in really cold weather.

How does a family party of geese or swans pick the direction in which to fly, and does a particular member lead the way? The intention to be on the wing is signalled to the group by quite conspicuous movements of the neck and head, and by special loud calls. In most ducks, migration direction and distance are apparently programmed instinctively. In the geese and swans, traditional knowledge of the breeding and non-breeding grounds is acquired the first time that the journeys are made in the company of experienced birds and repeated in subsequent years. Research has not yet established who leads the way, and whether the leaders change place every so often (though this seems likely).

It probably takes the group some time to reach their final cruising altitude and, in the case of swans, this has been shown to be as high as 27,000 feet (8,200 m), only 2,000 feet less than the height of Mount Everest. The birds can move at the edge of the jet stream, and thus Whooper Swans have been estimated to complete the autumn journey from Iceland to Ireland in as short a time as seven hours. They fly and breathe in what seem to us incredibly low temperatures — as low as −48°C. Strong cross winds can cause the flocks to

Facing page. Shelduck pair *Tadorna tadorna* over the Ribble Estuary where they nest in rabbit burrows.

Above. The Brent Goose *Branta bernicla* travels to nest at 80°N in parts of its range, further north than any other goose.

Right. The Whooper Swan *Cygnus cygnus* can fly to the British Isles from Iceland in half a day.

be displaced and thunderstorms are occasionally severe enough to cause mortality. Such an accident occurred in January 1978 over Norfolk when 140 Pinkfeet, Greylag *Anser anser*, Brent and Bean Geese were killed apparently while flying in a storm. Post-mortem examination showed ruptured livers and haemorrhaging consistent with damage by decompression or blast. The Bar-headed Goose is one of the most remarkable migrators, since the flocks fly twice a year over the high Himalayan range between India and Tibet. There is a record of 17 geese being photographed flying against the sun at Dehra Dun, India, at a height variously estimated at between 25,000 and 58,000 feet (7,620–17,680m). However, Lesser Snow Geese may fly at altitudes of up to 6,000 m over their North American flyway, and Brent Geese regularly cross the Greenland icecap which rises in places to 3,500 m.

The structure of the wing does not vary much in wildfowl, except that high altitude migrants, such as Bar-headed Geese, seem to have longer, thinner wings than other geese, and a few rather small-bodied migratory birds with white plumage, such as Snow Geese and Coscoroba Swans, have black-tipped primaries and secondaries for strength. The strengthening is provided by the pigment melanin, and is commonly found in the wing tips of pale-coloured birds that fly a great deal, such as gannets, gulls, terns and flamingos. It is an interesting observation that the Black Swan, which is not a regular migrant, manages with black wings and white-tipped primaries. And of course most of the larger migratory swans are all white.

In the equatorial and near tropical regions, wildfowl tend to show random movements in response to irregular, non-seasonal changes in rainfall amounts. Such dispersals are common, for instance, amongst Australian species such as the Freckled Duck, the Magpie Goose and the Black Swan. There is little if any breeding in years of severe drought.

Left. The Egyptian Goose *Alopochen aegyptiacus*, like all wildfowl, has eleven primary feathers in the wing (the eleventh is reduced). This bird was photographed in Amboseli National Park, Kenya.

Overleaf. Snow Geese *Anser caerulescens* have longer, narrower wings than Egyptian Geese. The black colour gives strength to the feathers.

CHAPTER 7

Wildfowl and People

Man has long looked upon wildfowl as a delicious source of protein, and in pursuit of a good meal has collected their eggs and young, rounded up moulting adults, and trapped and shot the winter flocks. Today, shooting is more often for sport than for the pot, but doubtless the modern hunter relishes his feast, as well as his day out in beautiful wild places.

Decoys were a trapping method developed in coastal Holland: indeed, the word is Dutch: *ende* meaning duck and *kooi* a cage. A pond was dug, with a number of netted pipes leading from it, and a dog employed to appear from behind a series of overlapping screens running along the pipes. The curiosity of ducks is such that they will swim after the dog (or any other land-based predator) and are caught beneath the nets. The device was introduced into Britain in the sixteenth century, when the Dutch were employed in some numbers to help drain the fens. Tame 'call ducks', bred especially small so that they did not eat much food and often white so that they were conspicuous, were used by the decoyman to call in their wild relatives. The word decoy has changed its meaning somewhat during the intervening years: it is now someone or something that lures another into danger, although it is also likely to be a beautifully carved and painted model. In their cruder versions, and floating on a pool near a flightpath, such models 'decoy' flying ducks and geese within reach of the guns.

Below. Nacton Decoy in Suffolk with Tom Baker, the decoyman, and his dog 'Piper'.

Facing page. The nomadic Australian Pink-eared Duck *Malacorhynchus membranaceus*.

Wildfowl have been put to many other uses. Their bones have been fashioned into whistles, for instance, and their flight feathers into pens (the American Declaration of Independence was signed with a goose quill); a favourite practice of Egyptian scarab-makers was (perhaps still is) artificially to age their forgeries by making a goose swallow them. Wildfowl featherdown is immensely valuable as an insulating material and has yet to be matched in quality by man-made substitutes. The collection of down from the nests of Eiders was a common practice among Eskimos and, in Iceland, the wild Eider Duck is farmed. In the early 1960s there were about 200 Eider farms in Iceland holding some 250,000 nesting females, each producing an average 19 g. of cleaned down. The down is collected at least twice; the first, taken early in incubation, is the best and least likely to need cleaning, the final collection being made after the eggs have hatched. The colonies are protected from Ravens *Corvus corax*, gulls, Arctic

Foxes and, recently, from Minks *Mustela vison*, and many farmers provide stone cubicles as nesting shelters.

Geese and ducks were among the earliest birds to be domesticated, the Greylag Goose about 5,000 years ago and the Mallard some centuries later (mammals had been domesticated much earlier — goats and sheep as long ago as 10,000 BC). In eastern Asia the Swan Goose *Anser cygnoides* and in central America the Muscovy were also brought into the farm, although we are not certain when. They were farmed for utilitarian purposes, as pets, and white ones were selected as sacrificial offerings. It has been said that all the important domesticated birds are seed-eaters or grazers which would have had an early association with Man through their raids on crops. The eastern race *rubirostris* of the Greylag is likely to have been the ancestor of most farmyard geese: it breeds further south than the western subspecies *anser* and therefore would have been in closer contact with early Man. It is large and kept

Below. The domestic farmyard goose *Anser anser* is descended from the eastern race of the Greylag. Well-developed gonads, associated with laying numerous eggs, produce the 'dropped tummy' effect.

Facing page. The marvellously soft down of the Eider Duck *Somateria mollissima* insulates the eggs and camouflages the nest.

Migratory Eastern Greylags, ducks and gallinules at Bharatpur Nature Reserve, India. This is the race. *Anser a. rubirostris*, of the Greylag that is the ancestor of the domestic goose on the previous page.

mainly for meat (the Romans discovered that it could be force-fed and fattened for its liver) but also for feathers and down, white geese having been quickly developed. Why are white forms so common in domestication? Feathers without pigment occur in most wild birds but the individuals seldom survive for long: the white feathers are not so strong and the birds are more easily picked off by predators because they stand out. I have already mentioned the religious significance of white offerings to the gods, but there is another important feature that seems to be associated with whiteness: the birds grow more quickly and convert food better than do the darker forms. For instance, the white phase of

the Lesser Snow Goose hatches half a day earlier and fledges faster than the blue phase. Why then are not all forms of domestic bird white? Because the dark forms lay earlier in the year and, as they lay for longer, produce more eggs. Early Man had the choice, eggs from dark birds or meat from white ones.

The Greylag was the ancestor of the Embden and Toulouse plus many other breeds including the Old English Grey Goose which was present when the Romans arrived in Britain — the Celts are said to have had a special reverence for it. The widespread custom of eating a goose on the Feast of St Michael (29 September) or, especially on the Continent, at Martinmass (11 November) probably has its roots in pre-Christian ritual when a bird was sacrificed to increase crop fertility: 'who so eats goose on Michaelmas Day shall never lack money his debts to pay'. Goose Fairs were held in many parts of the country. Nottingham Goose Fair, which is mentioned in a charter of 1284, starts now on the first Thursday in October but was originally a September event and became the principal hiring fair and autumn market of the midlands. During medieval times, 20,000 geese changed hands at a single fair; they were walked in along the ancient Fosse Way to the Old Market Square, through the Goose Gate on the eastern side of the town.

Two domestic breeds descend from the Swan Goose, the Chinese and the African. Their flesh is less fat and, as they are more tolerant of warm climates, they are usually the geese kept in tropical countries. The Mallard has produced five farmyard varieties that are popular for commercial use: the Aylesbury, Pekin and Rouen were selected for meat, the first two being white, and the Khaki Campbell and brown Indian Runner are excellent egg-laying strains. A flock of nine Pekins was imported into the USA in 1873 by a clipper ship captain, and the breed gained wide popularity as a table bird. About eight million are eaten in Britain annually and,

interestingly, we export large numbers of ducks' tongues and feet to China, where they are considered delicacies.

The Khaki Campbell was developed by Mrs Campbell, the wife of the village doctor of Uley, near Slimbridge in Gloucestershire (incidentally, 'khaki' is the Hindustani word for the colour of dust). The breed became increasingly popular during the 1920s and 30s with remarkable individual laying performances, such as 291 eggs without a pause and a total of 333 in 336 days.

The last of the four species of wildfowl to be domesticated was the Muscovy. It is related to the Mandarin and Carolina (or North American Wood Duck) and is a

perching duck of the rain forests. It was imported into Europe in 1550, and it is not certain why it was called the Muscovy — perhaps after the Muscovite Company that traded to South America, or after the Muysca Indians from whom it may have been obtained first. They had changed the bird relatively little: the domesticated form is larger than the wild species and occurs in a variety of dark green, grey and white colours, but there are no recognised breeds. It is clear that most of our farm animals were developed in the Old World, while New World Man was responsible for many of our agricultural plants: beans, potatoes, maize, peanuts,

Facing page. Drake Mallard *Anas platyrhynchos* wing-stretching. The birds are in first plumage.

Below. Domestic male Muscovy Duck *Cairina moschata*.

tomatoes, squash, avocados, manioc, tobacco, sweet peppers, pineapples and cocoa were all domesticated by South American Indians and, in many cases, they are greatly altered from the original wild type. Only the dog, whose domestication pre-dated all others, plant or animal, was common to both hemispheres.

Mute Swans were farmed in Europe from the middle ages, cygnets being put into pits to fatten on grain for Christmas. 'Syngnettys' were served at the sumptuous wedding feast of King Henry IV in 1403. White youngsters were specially valuable as they could be skinned and the skins, with the attached white down, sold for trimmings and powder puffs. Perhaps the Polish Swan became so common on the Continent partly because it was bred selectively by Man for its white juveniles. In Britain, the Mute Swan has always had a special

Above. Drake Mandarin *Aix galericulata* in breeding dress. The bird is regarded with special affection by the Chinese, who gave a pair to bridal couples as a symbol of marital fidelity.

Left. Wintering Pinkfeet *Anser brachyrhynchus* fly against the Southport gas holder from fields in Lancashire.

Overleaf. Black Swans *Cygnus atratus* in New Zealand. Introduced in the early 1860s from Australia, they bred so well that they have been regarded as pests.

relationship with the monarchy. Edward IV enacted that, except for the sons of the King, only freeholders of land above a certain value could mark or have a royalty for swans; swans on land below this value could be seized for the King. Swan-upping, during which the swans are counted and the juveniles are pinioned and their bills marked, still takes place on the river Thames on the first Monday in August before the cygnets are flying and while the adults are in wing moult. The only royalties in swans still remaining are those granted to the Dyers and Vintners Companies.

In the northern regions, where farming methods are becoming increasingly intensive, we do hear of wildfowl, and especially geese, damaging crops. It is actually a worldwide problem, but fortunately important only locally. When rice was planted in parts of northern Australia, the Magpie Goose, whose normal food is marsh plants such as sedge, came to be regarded as a pest of the rice paddies. In England, Pink-footed Geese may take carrots from fields where, until recently, the carrots would have been harvested before the birds arrived from Iceland in September. Sometimes (perhaps more frequently) the damage is caused by an introduced wildfowl species such as the Canada Goose in Britain and New Zealand. Western Man has certainly taken some pains to redistribute features of his natural world; he took the familiar House Sparrow *Passer domesticus* with him to America and the Chaffinch *Fringilla coelebs*, among others, to New Zealand — and wildfowl, being useful for hunting as well as beautiful, went too. The Mute Swan from the European continent has done well in Rhode Island, the Black Swan succeeded in New Zealand until a storm in 1968 devastated its food supply, and the Ruddy Duck (after an escape from the Wildfowl Trust at Slimbridge in 1959) is now spreading quite rapidly throughout Britain.

Man has had further effects by putting hazards in the way of the wild birds. Spent lead pellets from shotgun cartridges, and fishermen's lead weights, are eaten by ducks and swans, perhaps in mistake for grit. The lead is retained in the gizzard, ground down by abrasion

131

Above. The Canada Goose *Branta canadensis* is a species introduced into Europe and New Zealand from its native North America.

Facing page. Young Whooper Swan *Cygnus cygnus* on its first flight from Iceland killed by collision with power lines in Scotland.

with the grit already there and, helped by digestive acids, is dissolved and taken into the bloodstream. Lead poisoning paralyses, and the bird may starve to death with a full crop which its gut muscles simply cannot deal with. Power lines are an increasingly unlovely intrusion into our countryside and in Britain, at least, collision with man-made objects was the most commonly reported cause of death among Mute Swans twenty years ago; today lead poisoning may be in first place. We know that oil spills on the Thames have led to the deaths of 243 swans; scoters and eiders, because they spend large parts of the year at sea, are especially at risk

from discharged oil. The pesticide carbophenothion, used on sown grain to combat Wheat Bulb Fly, has killed Greylag Geese in Scotland — but not the Canada Geese on which the toxicity of the substance was earlier tested. This incident suggests that no chemical may be considered safe until every animal has tried a little of it and survived.

Four wildfowl species have become extinct within the last hundred years: the Pink-headed Duck *Rhodonessa caryophyllacea*, Labrador Duck *Camptorhynchus labradorius*, Auckland Islands Merganser and the Crested Shelduck *Tadorna cristata*. What evidence there is does not suggest that Man had a great hand in their decline, with the possible exception of the merganser. Here the Maoris may have removed the bird from mainland New Zealand, and introduced predators and Museum collecting activities later delivered the *coup de grâce* to the population on the Auckland Islands.

Man is now more aware of the need to conserve his natural world. An International Conference held in Iran in 1971 produced the Ramsar Convention whereby countries 'recognising the interdependence of Man and his environment' undertook to protect their own major wetlands. The results have been encouraging. Reserves

135

have been designated and created (for instance, Bharatpur in India and the Ouse Washes in England) and protective legislation enacted. On the Farne Islands, the Eider has long received a measure of protection, perhaps because of its practical utility to Man. St Cuthbert, who died there in 687 AD, appears in addition to have felt an affection for the birds: even now they are called locally St Cuthbert's Doves (because of the drake's cooing voice), Cuddy Ducks, Culverts and Cudberduce.

When the numbers of a species fall very low in the wild, Man may try to hold back extinction by captive breeding. The Hawaiian Goose, down to less than 50 individuals in 1949, has been restored, after captive encouragement at Slimbridge and in Hawaii, to the highlands of its native home. Sir Peter Scott can take the credit for saving this bird from oblivion. Another Englishman, the thirteenth Earl of Derby, was an earlier 'green-fingered' aviculturist. He bred the first Hawaiian Geese in 1834 at Knowsley Hall not far from Martin

Mere and, what was equally unusual, published a full account of his success. He also bred for the first time (among others) the Orinoco Goose *Neochen jubata* (and had Edward Lear paint it) and the Ashy-headed Goose. When he died in 1851 his unique collection was dispersed; the auctioneer's catalogue included 70 home-bred Passenger Pigeons, four Carolina Parakeets and a pair of Quagga — all now extinct. Dillon Ripley of the Smithsonian Institute may claim some responsibility for the present widespread distribution of the Carolina in the USA and Canada. Captive breeding, large-scale releases and the provision of nest boxes greatly increased the fragmented population.

The erection of boxes can increase the use made of a habitat where the tidying away of all the dead trees has resulted in a shortage of natural holes and hollows for nesting. This was the case, for instance, with the Chestnut Teal *Anas castanea* in Australia, the Grey Teal *A. gibberifrons* in New Zealand, the Goldeneye in Sweden, Poland and USSR, and the Goosander in Britain.

Facing page. The Hawaiian Goose *Branta sandvicensis* was saved from extinction by captive breeding.

Below. The Earl of Derby first bred captive Ashy-headed Geese *Chloëphaga poliocephala* at Knowsley Hall before 1850.

137

Above. Mallard *Anas platyrhynchos* flocks in autumn will feed on harvest waste and roost in dykes between the fields.

Left. Mute Swans *Cygnus olor* in Hamburg, Germany, rely on Man for much of their food.

Preceding page. The Goosander *Mergus merganser* has been persecuted because it is said to compete with Man for fish.

Many people find enjoyment in an involvement with ducks. Leisure is one of the four functions of the Wildfowl Trust centres, along with Education, Research and Conservation. Children are still taken there, and to the town park, 'to feed the ducks'. I am reminded of two very different anecdotes: Peter Scott tells that one of his earliest memories was of feeding the Mute Swans on the Round Pond in Kensington Gardens — 'huge sedate birds around which sped innumerable ravenous Pochards just arrived from the far side of the North Sea'. And the more recent bizarre observations (reported in *The Times*) of an artist by the Serpentine: a woman took a female duck from her shopping bag and put it in the water to lure the drakes. As the males approached, the woman snatched them, wrung their necks and tossed them in the litter baskets. When the artist remonstrated, she rounded on him: 'There are too many drakes in the world,' she cried. 'They are raping the ducks and the owls and even the swans!' Unfortunately artificial feeding, especially during the summer, does seem to attract Mallard drakes when they would normally be dispersed, and the attentions of these males can be very persistent and even violent.

We can probably all remember 'Ducks are a-dabbling, uptails all' if not the rest of the poem. And the lines about the creation of the duck 'and God must be laughing still at the sound that came out of his bill' (which suggests that the poet was unaware that only the

141

female quacks). One of my favourites is a rhyme for tossing a child up and catching it:

> Gray goose and gander,
> Waft your wings together,
> And carry the good King's daughter
> Over the one-strand river.

Much has been printed in an effort to decide who that marvellous doyenne of the nursery, Mother Goose, was and even whether she existed at all. She seems to derive from *La Mere Oie* or *Fru Gosen* of France and German folklore, and to be at least as old as the seventeenth century.

Wildfowl are, of course, wonderful subjects for the artist. Dead ducks abound in a certain 'furniture picture' type of still-life piece. The protestant north of Europe produced many such kitchen interior paintings during the seventeenth and eighteenth centuries, with vast quantities of raw food, game, guns and dogs. More

Facing page. These Galapagos Pintail *Anas bahamensis galapagensis* were photographed during a tourist trip with Lars Eric Linblad in the S.S. *Romantica*.

Above. The handsome Bronze-winged Duck *Anas specularis* occurs on the slopes of the Andes in Chile and Argentina.

recently, the beauty of wild swans, geese and ducks has been demonstrated by Audubon, Gould and Lillifors and also by Peter Scott's continued brilliance. (It is a fact that the art of painting in *tempera* was developed through the use of goose egg yolk as the fixative.) That photography can be an equally good medium of portrayal is shown superbly by this book.

In legend, swans and geese have exercised a mysterious hold over northern Man's imaginings, partly because they disappeared every spring to return in the autumn with their young. The need to provide a rational explanation for migration before its significance was understood, and before the discovery of the northern regions where the birds actually laid eggs, was at the root of many colourful and dominant myths. Because Barnacle Geese were said to hatch from goose barnacles (not the common acorn barnacles, but the sort that have an apparent 'head' hanging down on a 'neck'),

the bird was fish and could be eaten in Ireland on Fridays and during Lent. 'Swan Maiden' legends of girls who were really swans, but could be captured and even married if one could steal their feathery cloaks, must have similar origins in the need to account for the movements of wild creatures.

Placenames may include a reference to the ducks, geese and swans that were commonly seen or kept in domestication. Of obvious relationship are Swanbourne,

Facing page. Goose or Ship's Barnacles *Lepas anatifera* hanging from a floating log.

Below. Introduced Canada Geese *Branta canadensis* in New Zealand fly over a landscape cleared of native bush and now used for agriculture.

Buckinghamshire ('stream frequented by swans'), and Swanmore, Hampshire ('lake frequented by swans'). Beware though — most other places with the same first element (such as Swanage and Swanscombe) are said in *The Concise Oxford Dictionary of English Placenames* to derive from a different meaning: swineherd, as in Swindon! The older '*elfetu*', which is the Norse word for swan, and probably refers to the Whooper rather than the Mute, is commoner and gives us Altham, Lancashire ('flat meadow on a swan stream'), Eldmire, North Yorkshire ('swan mere'), Elvet Hall, Durham ('swan island'), Elvetham, Hampshire (another 'flat meadow . . .'), Elterwater in the Lakes, and Iltney, Essex (another 'swan island'). Note that most of these names refer to towns and villages in the north and east of England, which were the areas of greatest Viking influence. I have watched Whooper Swans at both ends of their migration, at Elterwater, Cumbria, and at Alftafjordur in southeast Iceland; both places were named centuries ago but still have their most conspicuous inhabitants.

The goose is immortalised in Gaisgill, Cumbria ('wild goose valley'); Gosbeck, Suffolk ('goose stream'); Goscote, Leicestershire ('hut for geese'); Gosfield, Essex; Gosford, Devon; Gosport, Hampshire; Goosey, Oxfordshire ('goose island'), and Goswick, Northumberland ('goose farm').

Ducks are found in the names of Doughton, Gloucestershire ('duck farm'), and Dukinfield, Greater Manchester ('field frequented by duck'). Digbeth Street in Stow-on-the-Wold derives from Duck Bath Street and was

Above. At one time the Whooper Swan *Cygnus cygnus* nested fairly commonly in northern Scotland but Man was responsible for its demise.

Right. African Yellow-bills *Anas undulata* flying in front of papyrus.

Wildfowl and People

originally the site of the town pond. As with swans, the older word '*ened*' for duck is more commonly used in placenames: Andwell, Hampshire; Anmer, Norfolk; Enborne, Berkshire; Enford, Wiltshire; Enmore, Somerset; and Entwistle, Lancashire ('river fork frequented by ducks').

The names that we give to the objects around us are often very ancient. English has its roots in the Germanic languages and ultimately in Indo-European. The Sanskrit word for goose is *hamsa*, in Latin it is *anser* and in Spanish *ansar*. Some other European names are:

Czech	*Husa*
Portuguese	*Ganso*
Dutch and German	*Gans*
Irish	*Goss*
Anglo-Saxon	*Gos*
Scandinavian (& Icelandic)	*Gas*
Russian	*Gus*
Welsh	*Gwydd*

Another combination of names and languages relates to the swan:

Latin	*Olor*
Irish	*Ealadh* or *Eala*
Breton	*Alarc'h*
Welsh	*Alarch*
Cornish	*Elerch*
Northumberland & Yorkshire	*Elk* or *Ilke*

149

Wildfowl and People

This final chapter, on Wildfowl and People, has touched on many aspects of our relationship with these lovely birds. In anthropology they are as fascinating as in biology. Would they be so well known were they not so beautiful? Why were many great names in ornithology attracted to them: Jean Delacour, Konrad Lorenz and Peter Scott? I can answer for them only in my own words; wildfowl are, in much of their behaviour, sufficiently like us to appear comfortably familiar, and in their untamed ways, wild enough to seem purely magical.

Right. Africa's smallest duck, the Hottentot Teal *Anas punctata*, and Lesser Flamingos *Phoeniconaias minor* at Lake Nakuru.

Below. A pair of Shovelers *Anas clypeata* feeding together in late winter.

Index

Note: page numbers in *italics* denote illustrations